SCIENCE COMMUNICATION

Science Communication: The Basics is an accessible yet critical introduction to science communication, which is viewed as the social conversation around science. It addresses why science communication matters, examines the evolution of theories and practices and explains concepts, myths, misunderstandings and challenges.

Massimiano Bucchi and Brian Trench navigate the foundations and key themes of science communication through numerous vignettes, examples, cases and arguments. They provide annotated recommended reading and a Lexicon summarising the understandings and uses of key terms in the field. Revealing science communication as a collective process and part of daily life, topics covered include science communication as part of the culture and our understanding of ourselves and the world; the history of science communication and the development of 'modern science'; policy and theoretical approaches; the growth of the professional practice, formal education and research in the field; evolving platforms for science communication; and quality, trust and ethical awareness in science communication.

Science Communication: The Basics is designed to be a concise primer and essential reading for newcomers to the field, including staff in research and policy institutions, students of the natural, human and social sciences, and general readers curious about the ways science is presented and perceived in society.

Science Communication: The Basics is the third in a triptych of works on science communication from the two authors. The other two works are the *Routledge Handbook of Public Communication of Science and Technology*, first published in 2008 and now in its third edition (2021), and a four-volume anthology of readings, *The Public Communication of Science* (2016), also published by Routledge.

Massimiano Bucchi has been the editor-in-chief of the leading journal in the field, *Public Understanding of Science* (2016–2019). He is Professor of Science and Technology in Society and Director of Master SCICOMM at the University of Trento, Italy, and has been visiting professor in Asia, Europe, North America and Oceania. He is the author of several books, published in over twenty countries, and a frequent contributor to newspapers and broadcasting. Among his books in English: *Science in Society* (2004); *Geniuses, Heroes and Saints: The Nobel Prize and the Public Image of Science* (2025).

Brian Trench, formerly Head of the School of Communications at Dublin City University, Ireland, has been the President (2014–2021) of the Public Communication of Science and Technology (PCST) network and reviews editor of the journal *Public Understanding of Science* (2017–2025). He is a former journalist and has published and presented widely on science communication and related topics. He writes on science, radical and cultural history, including for the magazine, *History Ireland*.

The Basics Series

The Basics is a highly successful series of accessible guidebooks which provide an overview of the fundamental principles of a subject area in a jargon-free and undaunting format.

Intended for students approaching a subject for the first time, the books both introduce the essentials of a subject and provide an ideal springboard for further study. With over 50 titles spanning subjects from artificial intelligence (AI) to women's studies, *The Basics* are an ideal starting point for students seeking to understand a subject area.

Each text comes with recommendations for further study and gradually introduces the complexities and nuances within a subject.

DREAMS
DALE MATHERS AND CAROLA MATHERS

JEWISH ETHICS
GEOFFREY D. CLAUSSEN

MODERN ARCHITECTURE
GRAHAM LIVESEY

MICROECONOMICS
THOMAS R. SADLER

STATISTICAL ANALYSIS
CHRISTER THRANE

ANTHROPOLOGY OF REPRODUCTION
SALLIE HAN AND CECÍLIA TOMORI

SCIENCE COMMUNICATION
MASSIMIANO BUCCHI AND BRIAN TRENCH

For more information about this series, please visit: www.routledge.com/The-Basics/book-series/B

SCIENCE COMMUNICATION

THE BASICS

Massimiano Bucchi and Brian Trench

Routledge
Taylor & Francis Group

LONDON AND NEW YORK

Designed cover image: Getty Images

First published 2025
by Routledge
4 Park Square, Milton Park, Abingdon, Oxon OX14 4RN

and by Routledge
605 Third Avenue, New York, NY 10158

Routledge is an imprint of the Taylor & Francis Group, an informa business

© 2025 Massimiano Bucchi and Brian Trench

British Library Cataloguing-in-Publication Data
A catalogue record for this book is available from the British Library

ISBN: 978-1-032-64671-8 (hbk)
ISBN: 978-1-032-64672-5 (pbk)
ISBN: 978-1-032-64674-9 (ebk)

DOI: 10.4324/9781032646749

Typeset in Bembo
by KnowledgeWorks Global Ltd.

CONTENTS

FIGURES

PREFACE

Science communication has become a topic of increasing interest and attention for institutions, policymakers, media and citizens over recent years. This little book offers a critical introduction to the theories and practices of science communication, their historical changes and present challenges.

It is not, and it cannot be, a comprehensive overview of the field but rather an opportunity to start navigating across key themes. Readers may wish to analyse these in more depth according to their needs and interests. Throughout the text, we indicate where such further exploration might go and we have included at the end, together with a Lexicon that summarises the understandings and uses of key terms, a list of recommended readings.

This book is the third in a triptych of works on science communication. The other two are the Routledge Handbook of Public Communication of Science and Technology, first published in 2008 and now in its third edition (2021), and a four-volume anthology of readings, The Public Communication of Science (2016). The full table of contents of this second work is included in our recommended readings; readers may seek in these classic texts and in the essays gathered and revised in the Handbook's editions more systematic treatments of certain topics.

The book is introductory yet built around our own view of science communication, namely science communication as the social conversation around science. A key element and implication of this view is that science communication is part of our daily lives, a collective process deeply – albeit, sometimes invisibly – woven into the fabric of our activities, experiences and conversations. We start each chapter with a vignette of a scene – real or imagined – that indicates this ubiquitous presence of science communication, and we point to many more examples and their implications through the main text.

This book has grown from a 25-year conversation between the two authors and with colleagues, trainees and students in different contexts and locations around the world. Our participation together in the conferences and the scientific committee of the PCST Network, as well as our joint and parallel involvement in science communication education and training, have contributed in myriad ways to better questions and more stimulating disagreements. We are particularly grateful to Eliana Fattorini for her support in revising the manuscript.

We hope readers will continue this conversation and do so in ways that we cannot anticipate or that may even run against the concepts and reasoning presented here. We also hope they find it as enjoyable as we have done. Conversations – and science communication – never end.

Massimiano Bucchi, Brian Trench, September 2024
Campiglia Marittima, Italy and Wexford, Ireland

THE SOCIAL CONVERSATION AROUND SCIENCE

This morning, during your breakfast, you may have paid attention to the nutrition facts displayed on the label of food or beverages on your table. These labels – and "food claims" in particular (e.g. the claim that a yoghurt is healthy for your guts, or that green tea is a source of antioxidant) are small pieces of science communication, written on the basis of expert analysis and reports by national and international food agencies. Similar advices on the properties of certain food may have been given to you by your doctor, or by friends and acquaintances during personal conversations. TV commercials often praise the beneficial properties of toothpaste by showing people dressed in white coats, moving around laboratories, pointing to graphs that are supposed to confirm its "scientifically" proven effectiveness. Before, during or after breakfast, you may have browsed the daily news by listening to radio or television, glanced at the daily press or through your social media feeds.

Maybe not today, but certainly sometimes during the week, you are likely to encounter in this way some news related to science: new discoveries or results by physicists or astronomers, data which feed our concern for climate change, discussions about the implications of innovations like synthetic meat, announcements of an important prize given to scientists for a particular achievement. An image may have caught your attention either on television or online, like the first "image" of a black hole attracted huge attention on worldwide media some years ago.

On your way to work, you may have passed a street or square with the name of a scientist from the past (e.g. Marie Curie, Guglielmo Marconi, Satyendranath Bose). Walking or sitting in a bus, you may have listened to a podcast that relates, at least in part, to the social or human sciences (economics, psychology, sociology, archaeology): experts that interpret certain

DOI: 10.4324/9781032646749-1

aspects of human behaviour; comments about economic, social and political trends; discoveries about our ancient past and so on. At work, during coffee or lunch break, chats with colleagues may have touched upon news related to science and technology, e.g. a new initiative by entrepreneurs like Elon Musk in the area of space exploration, or chip implants into the brain of a patient to address neurological conditions.

Later today or this evening, you may spend some time reading a novel or watching a movie or fiction. It is not uncommon, particularly these days, that themes and characters of fiction are inspired by science and scientists. But even a simple detective series is likely to portray scientific expertise (e.g. forensic science) aiding investigations of crimes.

Further examples could be added, but you got the point: science communication is ubiquitous. We are encountering it in the everyday and everywhere, in different forms and situations. Some forms and situations are explicitly framed and labelled as pertaining to science (e.g. science news sections in newspapers, tv or radio programmes about science). Other forms and situations are not explicitly framed as about science and communication, but they bear on related aspects and can have consequences, for example, on our perception of science and scientists. We summarise the variety of science communication as "the social conversation around science".

We find science communication in music, film, literature, café chat, street-names and postage stamps, as well as in the more obvious contexts of science media, science and natural history museums and documentaries about science, technology and nature. The notion of science communication as "the social conversation around science" emphasises the variety, richness and diversity of our encounters with science communication. Science communication is "society talking about science" in a variety of ways, with society obviously including scientists and scientific institutions.

Diversity should not be understood just as a question of formats. The aims of the actors engaged in different science communication formats of situations may also be diverse. A research institution or university may provide a press release and some images – about a research result, a study, an experiment or an achievement by one of its scientists – as a means to raise the visibility and profile of itself and its activities, emphasise the importance of the work by its researchers, popularise theories and concepts and legitimise the funding that they receive every year from taxpayers' money. News about climate change or synthetic meat can be used by different stakeholders, such as activists or food producers, to mobilise, protest and seek to influence regulations. Dedicating a street to a scientist is usually a way to honour her memory and boast national pride in association with a scientist's fame. Literary or film fiction uses scientific content or mimics science practices to sustain credibility or simply build a good story. For viewers or readers, such content may work as a source of inspiration, spark a desire to know more about a science-related topic or provide just plain enjoyment. Food labels with scientific jargon are intended by producers to comply with policy regulations. They are rarely read and even more rarely fully understood; still, they communicate indirectly to us a sense of reassurance, i.e. the fact that scientific expertise has officially validated their safety and properties.

The social conversation around science is ongoing and unpredictable. News about scientific results, originally meant to promote the image of research actors and support social consensus on a certain research area, can feed mobilisation and opposition (this happened, for example, with research on genetically modified organisms, GMOs). Vice versa, mobilisation and protest on science issues can lead to new research activities and results in the area, for example, of environmental risks or rare health diseases. Normally, "real" science and research inspires or are included in fiction, but more and more frequently, the production needs of fiction demand and inspire new, *ad hoc* research. This has happened, for example, with the movie *Interstellar*, inspiring an original paper by future Nobel laureate Kip Thorne, and with the animated Disney movie *Frozen*, fostering novel mathematical understandings of snow shapes and surfaces (Kirby and Ockert, 2021).

The social conversation around science is an open and potentially never-ending process. A piece of research can inspire a work of fiction, but this work of fiction can in turn inspire young people to embrace a career in a certain science or technical field. There are also relevant examples of fiction actually inspiring the development of new fields. In 1942, University of Columbia chemistry student and science fiction writer Isaac Asimov introduced in a short science fiction story ("Runaround") the term "robotics" and the so-called laws of robotics. Asimov's story contributed to inspire the creation in 1961 of Unimate, the first industrial robotic arm, developed and marketed with success by Asimov's fellow-student and entrepreneur Joseph Engelberger and his business partner George Devol.

Images and visual elements often play a key role in social conversation around science, enabling exchanges and cross-fertilisation among different areas of culture and society. The trajectory of pictures of the dodo bird painted by the 17th-century Flemish painters Roelant and

Jan Savery about the time when the last living exemplar was seen in Mauritius demonstrates a kind of conversation over time. In the 19th century, mathematician and writer Charles Dodgson (pseudonym, Lewis Carroll) introduced the bird as a character in *Alice's Adventures in Wonderland* (1865). He was likely inspired by a Savery image of a dodo in an Oxford gallery in guiding his illustrator, John Tenniel. In the same period, naturalist Richard Owen (1866) was figuring out how to reconstruct the fossil remains of a dodo sent to him at the British Museum and used Roelant Savery's paintings as a source. Three years later, Owen acknowledged he had been misled by such paintings to represent the dodo as "squat and overly obese" (Hume, Cheke and McOran-Campbell, 2009, p. 45), but by then the image of the clumsy and funny dodo had stuck. In this conversation loop images in art are seen to influence science and literature and settle in popular culture (Bucchi and Canadelli, 2015).

The notion of science communication as a social conversation around science implies that science communication is not just about an efficient transfer of information from specialists to non-specialists for various policy purposes. Although some actors involved in science communication may view it as a functional tool to achieve organisational or political aims, e.g. acceptance of certain results or innovations, this is just one of the many possible angles and perspectives, and it has been extensively and critically discussed in the past decades. Science communication is more than that. It is part of contemporary culture and society. It matters to our understanding of ourselves and our place in the world. The notion of science communication as a social conversation does not neglect the political or policy relevance of communication exchanges or activities related to science but puts them in a broader context, recognising, among other things, the cultural value of science communication.

CONVERSATION IN ART AND CULTURE

The English term "conversation" (similar to those used in other languages: "conversation" in French; "conversación" in Spanish; "conversazione" in Italian) comes from the Latin term "cum – versari", meaning not just getting together and talking but revolving in circles, enjoying being together. In medieval literature classics, conversation is often synonymous with social relationships ("Senza conversazione o familiaritade impossibile è conoscer li uomini", Dante, *Convivio*, I-vi-10). During the 15th and 16th centuries, the

"Sacred Conversation" became a common painting format, representing the Virgin Mary and child enthroned, with saints by their sides (and sometimes those commissioning the work) and musicians engaged in quiet communication. In this genre of representation, the Virgin Mary acts as in-between, bridging and enabling communication between the divine and human dimensions. She was the first intermediary of the Holy Spirit, transmitting divine wisdom to the Apostles and enabling them to converse in diverse languages, including languages they did not know. A typical example is the Sacred Conversation in the Church of San Zaccaria in Venice, painted by Giovanni Bellini in 1505 (see Figure 1.1). Within an architectonic frame are Saint Peter, San Gerolamo, Santa Caterina and Santa Lucia, each of them holding symbols referring to their role or martyrdom. The baby Christ pose evokes the later resurrection. The egg, probably imitated from other conversations, symbolises maternity, Easter, and thus resurrection again. The vegetation is accurate and symbolic: figs and ivy. An angel below the throne plays the *lira da braccio*, an instrument from that age.

Figure 1.1   Viewing La Sacra Conversazione (Giovanni Bellini, 1505) in San Zaccaria Church, Venice, before a conversation in September 2023 about the picture and its resonances for other kinds of conversation. Photo: Venice International University, by kind permission.

Later on, as often happens, the format and term migrated to secular representations, but with a very different meaning. By the end of the 16th century, conversation had become established as one of the cornerstones of appropriate behaviour at court. The full subtitle of the classic *Galateo: The Rules of Polite Behaviour* by Giovanni Della Casa (published in 1558 and soon translated into other languages), written itself as a conversation between a young person and an older, more expert person, reads "about modes to be adopted, or avoided, in common conversation". Conversation pieces became a common format in 17th- and 18th-century painting (particularly Dutch and British), real-life scenes of family or friends, sometimes engaged in conversations or other activities (meals, hunting, playing music). In a classic essay, art critic Mario Praz (1971) lists the four characteristics of this format: "i) two or more identifiable people, appearing as themselves and not as types or fictitious characters; ii) a background which describes the habitat of a family or group; iii) action, a gesture signifying conversation or at least communication of some kind from at least some components of the group; iv) privacy (i.e. not a public or official function)". An example of this is *The Schutz Family and Their Friends on a Terrace* (1725) by Philip Mercier. British painter William Hogarth also contributed to the popularity of the genre.

THEORIES OF CONVERSATION

In our first presentation of the idea of science communication as the social conversation around science, we explored theories of conversation, as developed by social philosophers and communication theorists, among others (Bucchi and Trench, 2021). The following paragraphs draw directly on that exploration, but with updated examples. We noted in opening that key thinkers on the public like Dewey and Habermas regarded talk and conversation as basic ingredients of civil, democratic and public life. Dewey (1927) declared the decline of conversation "the problem of the public". Habermas identified, perhaps in an idealistic way, rational conversation among equal, free individuals as one of the key elements in shaping the concept of the *public sphere*:

> By the public sphere we mean first of all a realm of our social life in which something approaching public opinion can be formed. Access is guaranteed to all citizens. *A portion of the public sphere comes into being in every conversation in which private individuals assemble to form a public body.* They then behave neither like business or professional people

transacting private affairs, nor like members of a constitutional order subject to the legal constraints of a state bureaucracy. Citizens behave as a public body when they confer in an unrestricted fashion–that is, with the guarantee of freedom of assembly and association and the freedom to express and publish their opinions-about matters of general interest.

(Habermas, 1974 p. 49)

Communications theorist James Carey advocated strongly from the 1970s for conversation as fundamental to democracy and for journalism as facilitating that conversation. By the 1990s, communications scholar Michael Schudson was writing of the "obsession" with, and "romance" of, conversation, drawing attention to differences between rules-bound problem-solving conversation, which can be difficult and even boring, and sociable conversation that is "an end-in-itself, an aesthetic pleasure" (1997, p. 300). In turn, Finnish scholar Risto Kunelius (2001, p. 45) questioned Schudson's hard distinctions:

If we deem only certain kinds of conversations democratically virtuous, we run the risk of uprooting democratic interaction from its cultural settings, and glorifying something that is at the same time in great danger of becoming irrelevant and hollow ... the 'public pleasure' of the participants is an important (preliminary) piece of evidence against the categorical idea of the uncomfortable and dangerous nature of public conversations.

There may well be useful analogies to be drawn here with conversations on and around science and the common tendency to prioritise those virtuous ones that (aim to) provide learning over those pleasurable ones that (merely) entertain. Also relevant to science communication, with its frequently asymmetrical relations, is the observation of communications scholar John Durham Peters that "conversation is no more free of history, power, and control than any other form of communication" (2000). In other words, conversation can be manipulated and is not necessarily open and equitable. Many attributes can be a handicap to participation, including gender, educational level, ethnicity and language. It takes conscious action to address these imbalances and exclusions.

CONVERSATION AND CONVERSATIONS

Two related usages of conversation are in play in our conceptualisation of science communication: a mode of interactive communication that is set in contrast with dissemination or other hierarchical modes, and a concept that embraces all that is being said on a certain matter in society. Our inclusive definition of science communication not only validates activities such as science cafés and science comedy that are oriented to pleasure but also recognises as part of the wider practice of science communication the "spontaneous" use in popular culture of images and ideas from and related to science.

Irish musician Hozier heard a TED talk by astrophysicist Katie Mack as he was writing songs for a new album. In *No Plan*, one of the songs on the album, he cites her: "As Mack said, there will be darkness again". Many cross-referring tweets later, in 2019, they met after one of his concerts, and the conversation has continued since then by various means, including in Mack's 2020 book, *The End of Everything (Astrophysically Speaking)*, in which she cited Hozier, and in a late 2023 visit by Hozier and his band to the Perimeter Institute in Canada, where Mack has held a chair in astrophysics and science communication since 2022. Hozier dutifully posed for photos, adding his thanks for the visit to the squiggles on a blackboard – the basic material of theoretical physics – and engaged in conversations on "black holes and dark matter, quantum gravity and the Einstein Field Equations", according to Mack's tweet.

Mack's Perimeter Institute colleague, Distinguished Visiting Research Chair and theoretical physicist Carlo Rovelli, has also sustained conversations across cultural zones and across the ages around his research topics (notably quantum gravity), but also on science in general. He is a prolific performer in public talks, public interviews and other forms of dialogue, including in the Pew Charitable Trust's Conversations on Science and with young colleagues in the Perimeter Conversations. His many interlocutors have included comedian and TV presenter (and physics graduate) Dara O Briain and science-engaged novelist John Banville, who has reviewed Rovelli's books for various media and who wrote novels on Kepler, Newton and Copernicus in the first phase of his literary career. In an extended conversation in December 2023 with the YouTube channel Downstream (Novara Media), Rovelli roamed over his radical political activity as a student, Buddhism, Greek scientific discoveries, religious dogmas, and the life of Russian revolutionary, systems theorist and science fiction author Mikhail Bogdanov. It is fair to observe of Rovelli that his broad cultural interests and personal openness ease the conversation around quantum gravity.

In other conversations, science communication may be said to happen accidentally or, at least, without this being the stated intention of the participants. That has been the case, for example, in radio shows discussing new psychological research or new books on nature, and for artists who have long explored scientific ideas and their possible application to the arts. We can find science communication as a conversation where there are no science communicators, self-designated or not. Conversation also emphasises long-term continuity in science communication: *conversazione* (in the Italian form) was a widely used designation in the 19th century for public displays, demonstrations and explanations of current science mounted by scientific societies for the enlightenment and entertainment of their expanding publics.

A characteristic of conversation articulated in communication studies and philosophy is that it is unpredictable and open-ended; we are also adopting deliberately this characteristic. Franco-Moroccan philosopher Ali Benmakhlouf (2016) stresses this, drawing insights from *Alice's Adventures in Wonderland*, which features many false starts and misunderstandings in conversations. Ideas, information or images from and about science can spread widely as one conversation opens another; in the process, the ideas, information and ideas inevitably acquire new meanings. This process does not always or only depart from and return to science, its actors and its institutions; it swirls in society somewhat independently and with interruptions, and that is what we intend to capture with the preposition, *around,* in our definition of science communication as the social conversation around science.

Many meanings of conversation can be accommodated in this discussion, from the structured engagement across society, the sponsored consultation, to spontaneous, even unruly, café chat. And there are other pertinent usages too: the Internet-mediated flow of information between experts and publics as expressed in the online initiative, *The Conversation*; the notion of a national conversation, often deployed with a desire to shift public opinion in a certain direction. Political leaders declare that they seek to change the conversation or change the narrative, which amounts to the same thing. Governments organise national conversations on dementia and immigration. Inevitably, there has been a kick-back at this instrumentalisation of conversation. It has been argued these national conversations are "futile" (Morris, 2016): "the belief in a national conversation is a belief in positive outcomes, in correctives, in *shoulds*" and the call for such conversation is "a hollow imperative".

Conversation has been having a moment in the few years since we started talking about science communication in these terms. A new model of chair is advertised as a "conversation chair", unwittingly recalling that an inverted two-seat chair of the 19th century also had that name. A radio station promotes itself as providing "conversation that counts". Self-improvement "experts" advise clients on how to start and sustain conversations. Social interest groups make the appeal, "we need to have the conversation", often meaning that they wish to change the trend. Conversation is also an active research topic beyond its obvious place in communication theory and the study of inter-personal communication, for example through conversation analysis that looks at the fine detail of gestural as well as verbal communication. Cognitive scientist Shane O'Mara (2023) explores the role of conversation in creating culture or – in the words of his book's sub-title – shaping our worlds. His view of conversation extends from our internal chats with ourselves to collective memory, foundation legends, and the building of nations "as conversations with the like-minded first, the convincing of others, and conflict with others again" (p. 162). The author presents the propagation of culture as happening "through countless small and large interactions and conversations and demonstrations … from listening to the radio running in the background, from conversations among friends" (p. 145). This view of conversation and conversations creating culture can be applied to scientific culture and thus to science communication as the enabler of that culture.

The conversation we speak of is both singular – the social conversation – and plural – the dispersed conversations of communities and colleagues, including the behind-the-scenes conversations of scientists that come increasingly into public view through social networks. Scientists' cafeteria and corridor conversations with each other and their conversations with colleagues elsewhere and with students (see Figure 1.2) are constituent parts of the making of science, and they resonate with public chat. In this and many other ways, the various modes of conversation around science connect. In representing these modes graphically (see Figure 7.1 in Chapter 7), we draw attention to the assumptions underlying a chosen communicative action. What appeared in earlier versions as a fairly fixed triad of deficit, dialogue and participation is intended here to be seen as dynamic: the range of modes is continually growing, but not just in the direction of more participation or co-creation, as a "progressivist" point of view might suggest. In the COVID-19

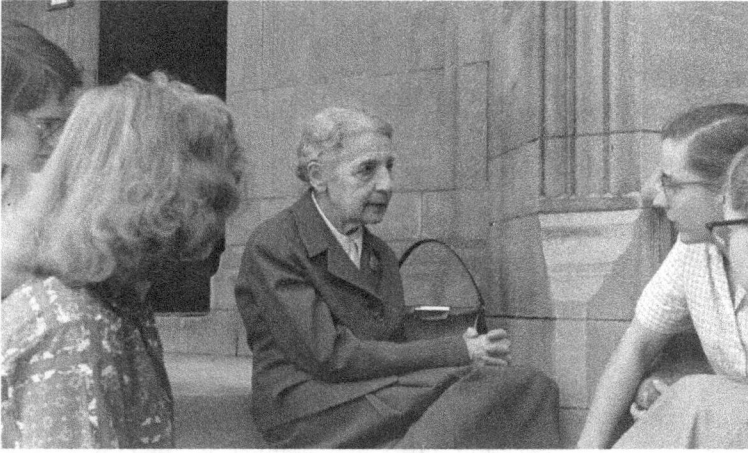

Figure 1.2   Lise Meitner, a key figure in 20th-century nuclear physics, with students at Bryn Mawr College, 1959. Photo by Heka Davis, courtesy of American Institute of Physics Emilio Segrè Visual Archives, Physics Today Collection.

crisis of 2020 and 2021, science was often invoked and scientific legitimacy was claimed by policy-makers to support measures that sought the population's compliance and thus limited social conversation. Science communication issues can travel across this spectrum in different ways and directions: an issue can initially be introduced in terms of dissemination and then move towards more participatory and even conflictual configurations (e.g. nuclear energy, GMOs). Vice versa, issues can be introduced and fostered into the social conversation by citizen mobilisation (e.g. specific health or environmental issues). Research activities and results can inspire – but also be inspired by – fictional content (e.g. robotics, space exploration).

The case of artificial intelligence (AI) is particularly interesting in this respect (Brause et al., 2023; Schäfer and Metag, 2021). Data from the Observa Science in Society monitor (a regular monitor of public perception and attitudes to science and technology in Italy that has been in place since 2003), for example, show that citizens recognise their general lack of information on AI, asking at the same time for rigid regulation. The issue is moving fluidly across the spectrum of social conversation, from a dissemination-like configuration (e.g. "what is AI?") to a more dialogic ("what are the implications for work and the economy?") and even controversial configuration (contested futures; regulation and governance, e.g. in connection with the temporary ban of ChatGPT in Italy in spring 2023, the British "summit" on AI policy and the temporary firing of OpenAI's chief executive officer in November 2023).

The perspective on science communication as social conversation/s that we outline here has implications for science communication research, both applied and fundamental, raising the priority given to questions of ethics, equity, inclusion, quality and history. It suggests that evaluation of science communication practice might be done in terms of how, and how much, a given practice or set of practices stimulates wider conversation. It also points to a wider context for analysis and reflection on science communication's social role and responsibility, putting long-standing issues of impact and effectiveness of science communication into a new context. It fosters reflection on the underlying values and purposes of science communication and on the largely tacit political and economic connotations of keywords like "responsible innovation" or of fashionable formats for presenting science to young audiences, as well as on their long-term consequences for the public perception and social role of science. A narrow definition of science communication has often carried with it a narrow definition of quality as impact or effectiveness, raising and reflecting expectations of quick fixes and solutions. Viewing science communication as social conversation expands and deepens also the quality challenge, increasing the range of relevant points of view and stakeholders: the quality of a conversation can never be judged just by one of the parties to that conversation.

Our idea of science communication as the social conversation around science has had some resonance in recent years. Reflecting on science communication as a contribution to democracy in a period of crisis, Davies (2022) draws on Priest's (2018) distinction between strategic and democratic purposes for science communication; Davies holds that it is possible to have "good faith engagement" while acknowledging honestly one's own position and the possibility of other positions. Even in an emergency such as the COVID-19 pandemic, Davies argues, that may mean accepting difference and diversity and embracing the unpredictability of science communication as the social conversation around science. Schikowitz and Davies (2024) revisited this concept as a frame for a study of Vienna housing activists' online communication,

suggesting that "to think of activists as science communicators is …
to move even further away from a model of science communication
as information transfer, towards framing it as public sense-making"
(p. 11) and that the social conversation context allows analysts of
science communication "to re-imagine our work as practitioners
in this way too, and to find innovative ways of contributing to such
conversations".

The social conversation concept has also been referenced in other
recent discussions of communication practices and models, including in relation to geosciences, climate change and disaster risks.
Exploring values in public representations of climate change, Fage-
Butler (2022) writes of the view of science communication as the
social conversation around science: "In this conceptualization, communicators of science are not understood in the narrow sense of
experts disseminating their scientific knowledge to the public; instead, science communication is conceived as being evident in the
heterogeneous, polyvocal and often lively societal discussions about
scientific topics".

Heterogeneous conversations are the stuff of culture. Musician
Nick Cave and writer Sean O' Hagan (2022) published their conversations around matters of life, faith and death. In his online publication, *Red Hand Files*, Cave offered a view of "good conversation"
that will stand well here as an encouragement to science communication conversations and is only incidentally referring to conversations on faith:

> A good [...] conversation begins with curiosity. It looks for common
> ground while making room for disagreement. It should be primarily
> about exchange of thoughts and information rather than instruction and
> it affords us, among other things, the great privilege of being wrong[1].

NOTE

1. https://www.theredhandfiles.com/to-speak-ones-mind/

REFERENCES

Benmakhlouf, A., 2016. *La conversation comme manière de vivre*. Paris: Albin Michel.
Brause, S. R., Zeng, J., Schäfer, M. S., Katzenbach, C., 2023. "Media
    Representations of Artificial Intelligence". In S. Lindgren (Ed.), *Handbook of
    Critical Studies of Artificial Intelligence*. Cheltenham: Edward Elgar Publishing.

Bucchi, M., Canadelli, E., 2015. *Nature immaginate: Immagini che hanno cambiato il nostro modo di vedere la natura*. Sansepolcro: Aboca Edizioni.

Bucchi, M., Trench, B., 2021. "Rethinking Science Communication as the Social Conversation Around Science". *JCOM* 20 (03), Y01. https://doi.org/10.22323/2.20030401

Cave, N., O'Hagan, S., 2022. *Faith, Hope and Carnage*. Edinburgh: Canongate Books.

Davies, S. R., 2022. "Science Communication at a Time of Crisis: Emergency, Democracy, and Persuasion". *Sustainability* 14 (9), 5103. https://doi.org/10.3390/su14095103

Dewey, J., 1927. *The Public and Its Problems*. New York: Henry Holt and Company.

Fage-Butler, A., 2022. "A Values-Based Approach to Knowledge in the Public's Representations of Climate Change on Social media". *Frontiers in Communication* 7:978670. https://doi.org/10.3389/fcomm.2022.978670

Habermas, J., 1974 [1964]. "The Public Sphere: An Encyclopedia Article (1964)". Trans. Lennox, S., Lennox, F. *New German Critique* 3, 49–55. https://doi.org/10.2307/487737 translation of entry in Fischer Lexicon (1964) *Staat und Politik*, Frankfurt am Main: Fischer.

Hume, J. P., Cheke, A. S., McOran-Campbell, A., 2009. "How Owen 'Stole' the Dodo: Academic Rivalry and Disputed Rights to a Newly-Discovered Subfossil Deposit in Nineteenth Century Mauritius". *Historical Biology* 21 (1–2), 33–49. https://doi.org/10.1080/08912960903101868

Kirby, D. A., Ockert, I., 2021. "Science and Technology in Film: Themes and Representations". In M. Bucchi, B. Trench (Eds.), *Handbook of Public Communication of Science and Technology*. London and New York: Routledge.

Kunelius, R., 2001. "Conversation: A Metaphor and a Method for Better Journalism?" *Journalism Studies* 2 (1), 31–54. https://doi.org/10.1080/14616700117091

Morris, W., 2016. Why Calls for a 'National Conversation' Are Futile. *The New York Times*, 2 August. https://www.nytimes.com/2016/08/07/magazine/why-calls-for-a-national-conversation-are-futile.html

O'Mara, S., 2023. *Talking Heads: The New Science of How Conversation Shapes Our Worlds*. London: The Bodley Head.

Owen, R., 1866. *Memoir on the Dodo*. London: Taylor and Francis.

Peters, J. D., 2000. *Speaking into the Air: A History of the Idea of Communication*. Chicago: University of Chicago Press.

Praz, M., 1971. *Conversation Pieces: A Survey of the Informal Group Portrait in Europe and America*. University Park: The Pennsylvania State University Press.

Priest, S., 2018. "Communicating Climate Change and Other Evidence-Based Controversies". In S. Priest, J. Goodwin, M. F. Dahlstrom (Eds.), *Ethics and Practice in Science Communication*. Chicago: The University of Chicago Press, 54–73.

Schäfer, M. S., Metag, J., 2021. "Audiences of Science Communication between Pluralisation, Fragmentation and Polarisation". In M. Bucchi, B. Trench (Eds.), *Routledge Handbook of Public Communication of Science and Technology*. London: Routledge, 291–304.

Schikowitz, A., Davies, S. R., 2024. "Housing Activists' Science Communication: Online Practices as Contextual and Reflexive". *JCOM* 23 (05), A01. https://doi.org/10.22323/2.23050201

Schudson, M., 1997. "Why Conversation Is Not the Soul of Democracy". *Critical Studies in Mass Communication* 14 (4), 297–309. https://doi.org/10.1080/15295039709367020.

2

SCIENCE COMMUNICATION BEFORE THERE WAS "SCIENCE COMMUNICATION"

When physicist John Tyndall accepted in 1872 a long-standing invitation to lecture in the United States, a New York publisher brought out his Fragments of Science for Unscientific People, selling out its first printing on the day it was released. That book went through many reprints and editions, including eight in Britain during Tyndall's lifetime. In the 1860s his book on sound was published in several editions; a book on water in clouds had 12 editions between 1872 and 1897; Lectures on Light had five editions in the same period. On Tyndall's US tour enthusiastic audiences, sometimes more than 1,000, heard his illustrated lectures on light. The Daily Tribune's verbatim accounts of his six New York lectures sold a total of 300,000 copies. Tyndall earned handsome reward for his US lectures, receiving the highest honoraria paid till then. In his lifetime, Tyndall was praised both for the "scope of his [scientific] investigations" and for "bringing the democracy into touch with scientific research ... In the promotion of the great scientific movement of the past 50 years he has played a part second to none". (Nature, 1887, pp. 217–8)

Setting out any history involves difficult decisions about where to start. There is a version of the history of science communication that spans a few decades of the later 20th and early 21st centuries, when the phrase, 'science communication', began to be widely used. Another version reaches back to the mid–1900s and the emergence of television science and interactive science centres. More recently, the advocacy for diversity and inclusion in science communication, as in other social and cultural sectors, has led to an embrace of the

DOI: 10.4324/9781032646749-2

knowledge of pre-literate and indigenous communities as science and thus of their inscriptions, visualisations and narratives as a form of science communication.

The intensifying contestation of the term 'scientific revolution' is underpinned by explorations of the scientific and engineering efforts of Asian and other non-European societies millennia before the present. Complex irrigation schemes, systematic lunar observations, indigenous medical practice and sophisticated mathematics are among the examples given of science before there was 'science'. The extraordinary feats of engineering by the builders of vast temples, pyramids, burial tumuli, stone circles and cities one, two and three millennia ago are seen by some as examples of early science. Many of these were based on detailed calculations, shared knowledge across the decades it took to build them and, in some cases, astronomical observations. Studying the stars and planets was also key to the remarkable achievements of Polynesian peoples navigating the Pacific Ocean. Such study was formalised and institutionalised in the construction of observatories in India and the development of models of the cosmos in ancient Greece and in Arab centres of learning.

In some such presentations, it is taken as axiomatic that where there was science, there was – or had to be – science communication. Building temples, amphitheatres, aqueducts and even cities may have taken many years or decades, so the guiding knowledge needed to be shared and passed on between generations. This may have happened through story-telling, demonstration and various kinds of inscription; our knowledge of such practices is greatly limited by the difficulties of studying mainly oral cultures and extinct languages of centuries ago.

That difficulty is a challenge and not a reason to reject the case for pre-literate science communication. Indeed, our own view of contemporary science communication is intended as encouragement to consider spoken and non-verbal communication as part of the bigger picture. An inclusive view of science and science communication in the contemporary context should, it would be thought, be associated with an inclusive view of historical science and science communication. As Orthia (2020) puts it, "if we are to (critically, cautiously) apply the word 'science' to knowledges beyond the recent West, we must consider opening 'science communication' to practices beyond it too". Orthia argues that dominant views of both terms are Eurocentric, tied to relatively recent history and exclusionary, but also that it would be "anachronistic and/or culturally imperialistic to label all the communication of knowledge across the world before 1800 as 'science communication'". Finlay et al. (2021) present examples of science communication from First Peoples and communities in 'global south' communities – in Australia, West Africa and India – as cases that "white, Western, European and Anglophone science communication can learn from".

The commitment to decolonising studies of cultures, history and knowledge is gathering force, supported by the current concern with diversity, equality and inclusivity in public life generally. These efforts are valid and necessary, and we acknowledge the risk in our own work of perpetuating Eurocentric bias. But we feel in historical and sociological terms justified in concentrating on so-called modern science and its communication because a defining and explicitly declared characteristic of that science was its formal communication among peers beyond those that generated the knowledge and, fairly immediately after that, the construction of a public – later, publics – beyond those peers again with whom the new knowledge could and should be shared. There is a significant historical break in the process commonly referred to as the "scientific revolution": a radical transformation in the forms of creation, validation and dissemination of knowledge between the 16th and 17th centuries in Europe.

Perhaps the most significant innovations brought by the latter to styles of thought and inquiry into nature were the following:

a.   the adoption of distinctive methods and procedures for scientific activity, primarily experimentation;
b.   the non-hierarchical character of knowledge: scholars were no longer bound to accept 'by fiat' what their predecessors had produced; instead, they were encouraged to analyse it directly on their own;
c.   the demise of a teleological, man-centred cosmology and extensive discussion of the most appropriate methods with which to study nature;
d.   the importance given to communication and the exchange of results and hypotheses – as opposed to the secrecy with which magical and alchemical works were shrouded – and the formation of a 'scientific community' with specific arenas for discussion.

This is not to imply that all the ideas and the practical and conceptual tools employed were radically new, even within Europe: anticipations of atomic theory or of heliocentrism, for instance, can be traced back to Ancient Greece (Butterfield, 1957). However, it was with the scientific revolution that these concepts, to a large extent, became the shared heritage of educated social groups. A global history of science (Poskett, 2022) argues that much that was considered novel in 17th century European science had precursors in Asia and Africa. The author's purpose is not so much to recall the glory days of Islamic science or Chinese inventions of a millennium ago but to trace the extensive roots outside Europe of the scientific revolution that has come to be seen as uniquely European.

The global view is compatible with the view that modern science as it developed in Europe was characterised by the systematic production, organisation and *distribution* of knowledge about the natural world subject to experimental proof. Communication was at the core of this process. No longer was valuable knowledge validated largely by reference to authority, be it religious authority or the consolidated philosophical tradition.

The founding charter of the Royal Society for the Improvement of Natural Knowledge by Experiment specified that the society was to consist of a president and other officers that included "Operator, Printer, Graver". The printing press was a foundation stone of the new scientific structures and those producing the society's publications and the visual images in them were key players in the scientific community. Leading scientists of the day worked on inventions to improve the engraving process to make better plates for publications. In an early example of visual science communication, Vesalius's anatomy manual *De Humani Corporis Fabrica* (1543) had invited students and fellows to perform the dissections and observations themselves, rather than relying on the traditional anatomical knowledge of the past. A crowning achievement of scientific illustration and also of microscopy was Robert Hooke's *Micrographia* (1665) with its painstaking rendering of life forms ("minute bodies" in the book's full title) as seen under a microscope and of the moon's surface as seen through a telescope.

Hooke, as Curator of Experiments at the Royal Society, produced the book on the request of King Charles II, patron of the society, who took a direct interest in scientific work and had been impressed by Christopher Wren's drawings of a louse, a flea and the wing of a fly. As part of his duties in the Royal Society, Hooke produced drawings for meetings of the Society's fellows; this activity extended then to providing information and entertainment for interested amateurs. *Micrographia* was a slower mover in the new book market than

Robert Boyle's earlier works on natural-philosophical experiments that were illustrated by Hooke. But Hooke's images for *Micrographia* had a large and long-lasting impact, being reproduced over the following decades and even centuries.

An underlying principle of the developing intellectual community of the 17th century was that knowledge could not be valued until it was circulated and communicated. For the new circles of 'natural philosophers', what mattered was priority, i.e., being the first to mark a discovery or obtain new empirical results. In order to claim priority and be recognised for their discovery, they had to communicate it. Initially, they did this through letters to colleagues and through books. The early years of modern science are marked by strident debates over priority, perhaps most famously, between Isaac Newton in England and Gottfried Leibniz in Germany over the invention of calculus.

A new medium was invented that made it easier to establish priorities and circulate new knowledge: the scientific journal. *Philosophical Transactions of the Royal Society* presented new results and discoveries proposed by scholars. Authority and tradition were displaced as key devices to evaluate knowledge. To decide what was new, reliable and worthy to print, another innovation was needed: peer review. The motto of the Royal Society was *nullius in verba*: we do not take anyone's propositions at face value; every fellow has the right and the duty to look into the results directly and judge for himself. Proofs and evidence needed to be demonstrated and witnessed, and doing this in the first instance for the judgement of peers opened up the possibility of creating a wider public for science. Science historian Steven Shapin (1990) observes that Boyle and his philosophical mentor Francis Bacon considered the lack of a public presence in alchemy as a sign that the practice was not scientific. They sought to develop a style of writing and discoursing that would facilitate public comprehension.

An interesting example of proto-popularisation of science is *Somnium, seu opus posthumum de astronomia lunari*, a short novel written in Latin in 1608 by astronomer and mathematician Johannes Kepler (but published only in 1634 by Kepler's son, Ludwig Kepler, several years after the death of his father). The book was written by Kepler to present the arguments of heliocentrism in a more accessible way by imagining a journey to the Moon, suggesting that an observer on the Moon would find the planet's movements as clearly visible as the Moon's activity is to the Earth's inhabitants. Well known to science fiction writers like Jules Verne and HG Wells, the book was later reappraised by scientists and popularisers like Carl Sagan as the first historical example of both science popularisation and science fiction.

The public for early natural philosophers was largely restricted to the educated, moneyed elite, including, most importantly, the patrons of science. The community of scientists (not known by this name until the 19th century) and the community of those interested in and supporting science were socially contiguous. When de Fontenelle published his *Entretiens sur la pluralité des mondes* (1686), he presented it as a series of conversations between a philosopher and a marquise. From our own perspective on science communication as the social conversation around science, we might consider de Fontenelle's work as noteworthy in adopting the format of conversation between specialists and non-specialists as a means to make new knowledge more widely known. But de Fontenelle declared his intention was to deal with 'philosophy' – by which he meant astronomy and physics, mainly – "in the least philosophical manner possible", acceptable both to the gentry and the scientists. He advises the marquise, "Let us content ourselves with being a select little band and not disclose our mysteries to the people" (cited in Goldsmith, 1986, p. 4).

A near-contemporary of de Fontenelle, Marie Meurdrac, sought to explain and promote chemistry to women in *La Chymie charitable et facile, en faveur des dames* (1674). From the early years of the 18th century, the emerging narrative genre of science for lay people often targeted female readers. *The Ladies' Diary*, established in 1704, was an outlet for scientific information, puzzles and medical advice (see Figure 2.1). It aimed to present "new improvements in arts and sciences and many entertaining particulars – designed for the use and version of the fair sex". Women were also seen in this context as 'symbols of ignorance, goodwill and curiosity' (Raichvarg and Jacques, 1991, p. 39); examples of works by men aimed at women are Algarotti's classic *Newtonianism for Ladies* (1739) and de Lalande's *L'Astronomie des Dames* (1785).

In the 18th century, a new genre of public entertainment grew as itinerant lecturers presented scientific spectacles for consumption by a public of men and women, boys and girls. English painter Joseph Wright of Derby depicted such a mixed audience in his 1760s images of a natural philosopher lecturing on the use of the orrery (a mechanical representation of the solar system) and another demonstrating a version of the air pump that Royal Society co-founder

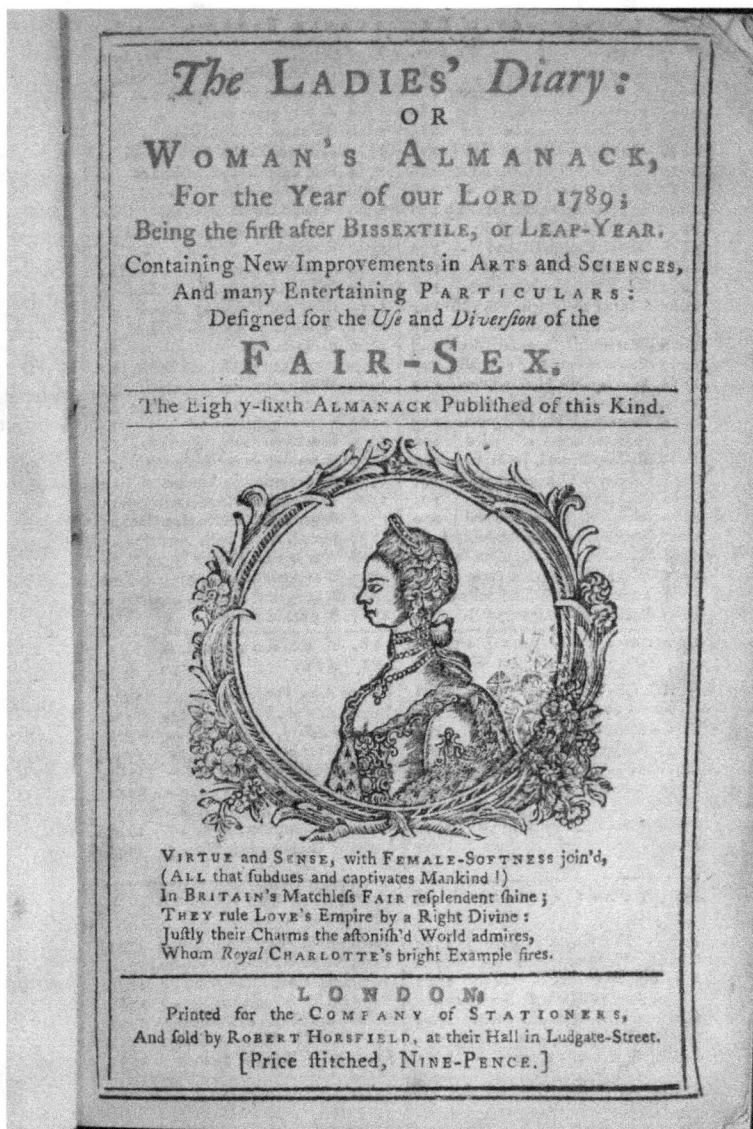

Figure 2.1 Cover of *The Ladies' Diary or the Woman's Almanack*, London. Image:
© Alamy/Museum: Private Collection.

Robert Boyle devised and demonstrated with the assistance of Robert Hooke in the 1660s.

In October 1743, public displays of more than fifty chemical experiments were presented in Dublin for 'The Amusement and Entertainment of Ladies, as well as Gentlemen'. The performer was Boyle Godfrey, a doctor, scientist and entrepreneur who had left his native London to avoid angry creditors and who announced himself to be a Professor of Chemistry. The experiments advertised on posters around the city were said to include some that had been presented to the King of Prussia and others at universities in Holland and Germany and at Oxford and Cambridge. As came to be typical of such displays, they involved smells, explosions and flames.

Blending science-based education and entertainment in this way also extended to magic tricks. Prussian-born Gustavus Katterfelto toured Britain from the 1770s to 1790s with shows under the title, Wonders! Wonders! Wonders!, mixing scientific content and instruments with quackery, conjuring and pure invention. He claimed to have invented a solar microscope, to have launched the first hot-air balloon and to be the greatest natural philosopher since Isaac Newton. His audiences extended to Britain's royal family.

Cabinets of curiosities were used from the 17th century as a means of sharing more widely objects of particular interest to science. These cabinets (rooms) in the houses and palaces of the upper classes and of the earliest scientific societies displayed animal skeletons, plant specimens and minerals, but also made articles such as religious relics, clockwork automata, sculptures and ceramics from communities. Much of this exotica was the bounty of colonial explorations and associated natural history expeditions. Some cabinets served mainly aesthetic and entertainment purposes, while others were more practical and scientific. They are considered to be precursors to the museums of natural history that began to be established in the 18th century and grew greatly in number and geographic spread in the 19th century.

A GOLDEN AGE OF SCIENCE COMMUNICATION

The 19th century has come to be seen as a 'golden age' of popularisation, the need and opportunity for which arose from the progressive differentiation of scientific and public culture (Shapin, 1990). The reasons for this included the increasing development of research activities, their progressive institutionalisation into distinct social and physical spaces (institutes, laboratories, universities), social recognition, growing secularisation of societies, particularly in certain parts of the world, and the distinctive appeal of scientific content and characters within broader, educated audiences.

Several contributors to a global survey of science communication's more recent development (Gascoigne et al., 2020) note that the history of science communication in their respective South American countries is often stated to start in the mid-19th century but then add that the documentation of this early phase is scant. Very considerable work has been done, however, on the development of the science popularisation 'marketplace' in Britain in the Victorian period. This historiography shows that through the course of the 19th century, the publics for science and the locales for popularisation increased and diversified.

Through scientific societies and academies a public developed of the educated elite who enjoyed lectures with demonstrations as "fashionable topics for polite conversation in elite society" (Finnegan, 2017). Humphry Davy was already a star when he was in his twenties, attracting big crowds to the Royal Institution and the Royal Society in London for his lectures. In 1808, Davy presented at the Royal Society his discovery of barium, calcium, boron, strontium, and magnesium, demonstrating his electrochemical methods, producing "sparks, explosions, and unusual odours, all guaranteed to excite the audience" (Kenyon, 2009). "Ladies in the audience twittered at Davy's fireworks and surreptitiously took notes. Aristocrats preened and even took turns standing in as Davy's assistant. He was revered by the audience as a scientific wunderkind". The enthusiastic audiences included the poet William Wordsworth, who came to hear the use of metaphor and analogy in Davy's carefully prepared and rehearsed lectures. When Davy delivered a series of lectures in Dublin in 1810, his audience included the novelist Maria Edgeworth. On the basis of strong ticket sales, the Dublin Society, who hosted Davy, was able to pay him fees for a week's work that were many times the value of his annual salary at the Royal Institution.

Jane Marcet (1769–1858) had a female readership in mind when she set out to summarise and popularise Davy's work in *Conversations on Chemistry, Intended More Especially for the Female Sex – In Which the Elements of that Science Are Familiarly Explained and Illustrated by Experiments*. Marcet moved in intellectual circles and attended Davy's lectures. She published her first edition anonymously in 1805 as one of the first elementary science textbooks. Jane Marcet's target readership was girls, so her conversations have a different character than those of Fontenelle more than a century earlier. Her work was used in women's colleges in the United States and France, as well as in England, where it was first published. Marcet applied the Conversations approach to books on political economy and natural philosophy. She also undertook popularisation efforts under the auspices of the Society for Diffusion of Useful Knowledge. Marcet was not explicitly identified as the author of *Conversations on Chemistry* until the 12th edition of 1832. Her last edition of *Conversations on Chemistry* appeared when she was 84.

Humphry Davy's successor at the Royal Institution, Michael Faraday, read Marcet's book as a young person and later recalled it as an important influence on him. He became an illustrious popular lecturer, initiating the Royal Institution's Christmas lectures in 1825, which he delivered 18 times over the following 30 years. Faraday conceived the lectures as being for a wide audience, including young people, for whom lectures came to be commonly presented at museums and other institutions (see Figure 2.2). The Royal Institution series has continued for two centuries, being continuously broadcast since 1936 and occasionally presented at locations outside Britain.

In 1826–1827, the German explorer and natural historian Alexander von Humboldt presented his views of the natural world informed by

Figure 2.2   Explorer Paul Du Chaillu Lecturing to the Young People of Boston, Harper's Weekly, 1869. Image: © CORBIS/Corbis via Getty Images.

his adventures and findings in South America to audiences numbering up to one thousand in Berlin. These eighty 'Cosmos-lectures' talks were considered the highpoint of culture in Berlin at that time; their content was the foundation of *Kosmos – Entwurf einer physischen Weltbeschreibung* (Cosmos – Design of a physical description of the world), prepared for publication in the late 1820s but delayed by the many demands on Humboldt's attention and his precarious health for 16 years. On its publication in the 1840s, it quickly acquired the status of a classic. Humboldt was a major public figure and was commemorated in many countries, including on postage stamps issued in Chile, Colombia, Cuba, Peru, Uruguay, Venezuela, Mozambique and the USSR, as well as in both parts of Germany before reunification.

As with Humboldt and many other cases, public talks on science with illustrations and demonstrations were the basis for popular books. John Tyndall, a successor to Davy at the Royal Institution, lectured extensively there and in other settings, including at working men's clubs and to audiences of many hundreds. Presenting prizes at the Mechanics' Institute in Preston, in north-west England, in 1885, he spoke on energy, evolution and germ theory to an audience of over 5,000, as reported in the *Leeds Mercury* (cited by Jackson, 2018). Tyndall's US lecture tour in 1872 had been a great success and highly profitable, though Tyndall used his fees to endow scholarships in physics. But in an early demonstration of what came to be known much later as the 'Sagan Effect', referring to the contrast between Carl Sagan's profile as a public figure and that as a scientist, Tyndall was deemed by some to have sacrificed his standing as a scientist for his prowess as a populariser.

Through their books and public lectures, some popularisers became public celebrities of their time. Paolo Mantegazza was a surgeon, a professor of anthropology and director of a museum of anthropology, a member of the Italian parliament, a novelist and a public lecturer on health and medicine, starting in the 1860s at his hospital in Milan. From 1865 to 1905, he published forty 'almanacs' on topics in medical science and promoting better hygiene and health. He advocated for state support of scientific research, saying that to secure that, "we [scientists] should go out of our laboratories and museums, we should become apostles of the message of science for us and for our children" (Mantegazza, 1881, cited in Turbil, 2017).

Mantegazza was a doctor and scientist with an aptitude for performance, but there were also in this period performers with an aptitude for science. Levie Kingsbergen Maju was a Dutch magician who reinvented himself as a science populariser in the 1860s. He appeared regularly on the lecture scene in the winter and in festivals during the summer, covering cities, towns and villages across the Netherlands, but also on a few occasions in Berlin and London. Having moved from conjuring to communicating science, he dropped his title of Professor of Sleight of Hand in favour of Member of the Polytechnic, milking his association with the Royal Polytechnic Institution of London, which also had its own programme of 'rational entertainment'. Maju borrowed displays and techniques from the London Polytechnic, presenting "the magnification of water droplets, living cheese mites and bacteria samples to appreciative Dutch audiences" (da Rocha Gonçalves, 2020), but adding his own microscopic slides of the development of a young salmon, which won the attention and approval of natural historians. He often delivered two presentations in one evening, the first for a better-off, fee-paying audience, and the second for "the common people, with free entrance and refreshments". From 1878, Maju presented a show centred on Edison's phonograph, recording himself playing the trumpet, then playing alongside the recorded music, and explaining principles of acoustics; he later incorporated a telephone into his performances.

The audiences for science had widened as new organisations dedicated to public education of the 'lower' classes were set up. The Mechanics' Institutes had started in the 1820s to provide education, particularly in science and arts (technology, in today's language), for the artisans, mechanics and operatives in the expanding manufacturing industries. "Scientific subjects became a form of 'shop talk' rather than a topic of interest to a social elite" (Finnegan, 2017). The British Association for the Advancement of Science, founded in 1831, created a platform for the presentation of current scientific knowledge that increasingly took on a public communication role. Other organisations promoting public science included the Society for Distribution of Useful Knowledge (1826–1846) and the Gilchrist Educational Trust which presented Science For All lectures in towns and cities around Britain and Ireland from the 1840s to the 1910s. Science For All was also the title of several publications, including a four-volume collection of articles about various aspects of science, published in the 1880s.

Science was also being made available to wider publics through events known as *conversazioni* (conversations, in Italian). These events could include talks, demonstrations, musical performances and exhibits. They were a common part of the programmes of scientific

Figure 2.3   Scientific Conversazione at Apothecaries' Hall, from The Illustrated London News, 1855. Image: © Getty Images/Universal History Archive.

conferences, societies and museums in the mid- and late-19th centuries. In many ways, they anticipated the 'open days' at research centres and universities in more recent times, but with an element also of contemporary science cafés (see Figure 2.3).

These *conversazioni* were well-enough established to earn inclusion in an 1862 guide to London's social life written by prolific author George Augustus Sala, who in his survey of the activities and entertainments in the city noted 11 p.m. as the hour of the scientific conversazione (Murphy, 2021). A lively and stylish conversazione was a required part of the annual meetings of the British Association for the Advancement of Science – an occasion for popularisation of and between the sciences – and host cities "competed with one another in contriving the most interesting gatherings" (Murphy, 2021, p. 90). An 1881 conversazione at the Royal Dublin Society included demonstrations of scientific instruments, short talks on current topics, orchestral music and a demonstration of the new means of relaying musical concerts over distances, the telephone.

Other elements of science popularisation's diversity in the mid- and late-19th century included natural history, other kinds of science museums, and the associated great exhibitions, notably those held in London, Chicago and Paris. The Science Museum in London opened in 1857, primarily in its first phase as "an educational institution, attempting to provide teaching in basic principles to teachers and skilled workers" (Hudson, 1988, cited by Schiele, 2021). Engineers were to the fore in the establishment of the Deutsches Museum in Munich in 1903; it opened its doors in 1906 to displays of 'masterpieces' of science and technology. The emphasis on technology was stronger again in museums opened in the early 20th century, such as the New York Museum of Science and Industry and the Palace of Discovery in Paris.

Discoveries, experiments and activities by visible scientists like Louis Pasteur in the late 19th century were prominent in both the national and international daily press (Bucchi, 1997). The transformation of Albert Einstein into a worldwide celebrity started with reports of the empirical confirmation of general relativity that made the front pages of newspapers like *The Times* of London and the *New York Times* in November 1919. Initially ignored by colleagues, the discovery of X-rays by Wilhem Roentgen in 1895 captured public attention after a first X ray picture of Roentgen's wife's left hand was published by Viennese newspaper *Die Presse* and then widely relaunched in the international press. During the 19th and 20th centuries, newspapers were a typical setting for scientists' engagement in public discussions that bore on science-related issues, both through interviews and comments or letters. In 1881, for example, in the context of a heated discussion about vivisection in Britain, the *Times* published a letter by Charles Darwin: while expressing his aversion to cruelty to animals, Darwin at the same time voiced concern that progress in physiology might be affected by restrictive legislation on the matter. Einstein was interviewed regularly by newspapers worldwide about science, politics, art, music, and occasionally contributed with his own writings.

Popular science magazines have a long tradition that goes back to the end of the 18th century, and it is initially strongly connected to the history of science publishing, when periodicals like *The Philosophical Magazine* (established in 1798) declared their mission to "give the public as early an account as possible of everything new or curious in the scientific world" (Hamblyn, 2001, p. 115). Along with the increasing specialisation of science publications and the growth of public interest in science, dedicated popular magazines began to flourish. *Scientific American*, for example, was founded in 1845. Popular science magazines appeared in significant numbers from the late 19th century, prominent among them *Popular Science Monthly*, which started in the United States in 1872 as a source of accessible information on science in general targeted at those who practiced science in particular. The magazine later moved to publishing shorter articles for wider audiences. The German magazine, *Natur*, started in 1852 with the ambition to reach all social strata and ran for fifty years. *National Geographic* was established in 1888, eventually winning a global audience. In France, popularisation magazine *La Nature* was founded in 1873 by chemist Gaston Tissandier.

Science popularisers in the late 19th century profited from changes in the publishing business and the increasing reading audience, but their success also testified to the increasing relevance of science as a cultural force. The sales figures of books like *Brewer's Guide to the Scientific Knowledge of Things Familiar* (published first in 1847 and selling 195,000 copies up to 1892) are impressive even by contemporary standards.

As indicated in the early mission of *Popular Science Monthly*, an important audience for popularisation in the late 19th century were scientists themselves, increasingly dispersed into separate communities through specialisation. Agnes Mary Clerke, self-educated in astronomy, made a career as a science writer, starting in the 1870s as a contributor to the *Edinburgh Review* of commentaries on trends in astronomy for those with a professional or highly developed interest in the field. Clerke was an early example of the "new order of worker" identified by her friend and colleague Margaret Huggins, "whose mission is to collect, collate, correlate, and digest the mass of observations and papers ... to prepare material for experts, and at the same time to inform and interest the general public" (1907). Clerke's *Popular History of Astronomy during the 19th Century* was published in 1885 to great acclaim, including from senior astronomers who appreciated its clear and concise presentation of their community's work; it was reprinted and republished in several editions from 1887 to 1902.

HG Wells, best-known for his science fiction, wrote his first published work as a text book on biology. He took a keen interest in scientific developments and defended popular science in *Nature* (Wells, 1894) against the possible criticisms of scientists, but also conceding that much popular science writing failed to find the appropriate language and was marred by clumsy attempts at humour or poor organisation of material. Greater respect should be shown to the reader, Wells wrote in a comment that echoes down the decades.

In addition to the flourishing of popularisation content in explicit terms, science became, particularly in the second half of the 19th century, a powerful cultural influence, inspiring writers and artists in broader terms. Wells is one of the leading examples of a popular writer deeply inspired not just by the content but by the scientific approach as well as by the impact of knowledge and technological developments in fields like biology or physics on everyday life, society and politics. After a modest beginning as a theatre playwright, French writer Jules Verne dedicated himself to novels inspired by the developments of science and by scientific characters, achieving international success with novels like *Journey to the Centre of the Earth* (1864) and *From the Earth to the Moon* (1865). In Italy, a sign of this popularity of science is offered by the title of the 1892 best seller, which is considered the cornerstone of Italian national cuisine, Pellegrino Artusi's *La Scienza in Cucina* (*Science in the Kitchen*).

Both Wells and Verne, with their works, contributed to inspiring further generations of scientists. Physicist Leo Szilard, who had a key role in suggesting an atomic programme to the government of the United States by signing the famous letter to President Roosevelt with his former professor Albert Einstein, first thought about nuclear weapons upon reading a 1914 book by Wells, *The World Set Free*. Similarly, pioneers of space flight and astronautics like Robert Goddard and Hermann Oberth were significantly influenced by Wells and Verne, whose works they read in their youth.

Still, until the early 20th century, science was more a vocation than a profession. Its increasing professionalisation and specialisation was one of the drivers of the perception that science had indeed become 'too complicated' and distant from everyday life, thus requiring a mediation (often referred to as 'translation') in order to be explained and disseminated. A generation of scientists radicalised by the experiences of the First World War – and not least by the misuse of science in the development of chemical and other mass-destructive weapons – gave new impetus to popular science in the 1920s and 1930s. Philosopher Bertrand Russell wrote accessible guides to current science such as *The ABC of Atoms* (1923), printed six times in the first six years of its existence; Russell later became a notable campaigner against the application of the science of atoms to warfare. Biologist J.B.S. Haldane kept up a steady flow of commentaries on contemporary science in the communist newspaper, *The Daily Worker*, and the *Manchester Guardian* during the 1930s. He was acknowledged by science fiction writer Arthur C. Clarke as "perhaps the most brilliant science populariser of his generation". Haldane (1941) reflected on the challenges of writing popular science articles, advising his colleagues to ask themselves: "For whom are you writing? This is even more important than the choice of subject" (p. 154) and advising them when they have written an article to "put it away for six months and see if you still understand it yourself" (p. 156). Haldane's rather high-minded notion was that "popular science can be of real value by emphasizing the unity of human knowledge and endeavour, at their best" (p. 158).

In 1937, engineer and artist Alexander Calder created a sculpture that continues to resonate – almost literally – as a piece of science, environmental, health and political communication nearly ninety years later. In conversation first with his Catalan artist friend Joan Miró and then with the Catalan architect Josep Lluís Sert, who was charged with designing the Spanish Republican government's exhibit for the Paris Exposition des Arts et Techniques Appliquées à la Vie Moderne, Calder developed a modern mercury fountain that still runs today. While recalling much older mercury fountains from Moorish times, it also commemorates generations of workers who died or were poisoned mining mercury in Almadén, south-west Spain, and salutes the republican fighters who defended the mines in the Spanish Civil War against General Franco's attempts to take them over. Based on complex geometry and physics, like so many of Calder's works, it is displayed in the Barcelona museum of the Joan Miró Foundation within a thick glass container to protect viewers from the toxic fumes. It demonstrates the remarkable physical properties and danger of the element Hg. It was a piece of artistic science communication long before 'science communication', and even longer before 'art-science'.

During the 20th century, the diffusion and popular success of mass media like cinema, radio and later television provided new platforms and opportunities for science communication and filtering of science topics and figures into popular culture. Science topics and science figures have been present in cinema since the very beginning of movie production in the early 20th century, for example, with film biographies dedicated to visible scientists like Louis Pasteur or Marie Curie or to discoveries like X-rays or evolution theory. A classic example is *The Story of Louis Pasteur* from 1936, directed by William Dieterle and greatly acclaimed at the time, receiving also three Academy Awards. Extensive references to science concepts have also been a regular feature of science fiction movies. Over several decades, selected scientists have had roles as advisers and consultants for movie productions. For the silent movie *Frau im Mond* (Woman in the Moon, 1929), German director Fritz Lang asked physicist and pioneer of space science Hermann Oberth to design a realistic model of a space rocket. Presenting for the first time a scientific context for space travel, the film made a very strong impression on audiences. Oberth used the funds received from Lang for his actual space rocket research and dedicated his book *Ways of Spaceflight* to Lang (1928; see Kirby, 2003).

Scientists also contributed to radio and TV programmes. In the early history of radio science, a standout moment came in 1948 when the British astronomer Fred Hoyle from 1948 gave several lectures on astronomy on BBC radio, including a series entitled The Nature of the Universe (1949–1950). During one of these programmes, Hoyle coined the term "big bang" to refer critically to a competing theory about the origin universe (Hoyle, with Thomas Gold and Hermann Bondi, had proposed a theory of a steady-state universe). The term immediately caught the attention of both specialists and the public, eventually becoming a synonym of the origin of the universe in popular discourse.

After World War II, changes in the global and policy landscape redefined the social and political role of science and its modes of funding and organisation. The term 'science policy' embodied the new political and social awareness that science could play a key role in terms of economic growth and political and military power, that it could be appropriately steered through funding mechanisms and other arrangements, and that its results could be used to orient political decisions.

The first use of nuclear bombs, resulting from a massive scientific effort through World War II, prompted redefinition of science popularisation in conceptual and ideological terms. The newly established Unesco agency of the United Nations explored the need for and challenges of science popularisation in this context, in terms of the importance of science in everyday life, the promotion of rational thinking among the population and the requirement that those in positions of power should understand science and its potential. English scientists and popularisers Julian Huxley and Joseph Needham were leading figures in Unesco and saw journalism as the main means to build appreciation of science. Danish journalist Borge Michelsen became the first Unesco's Division for Science and its Popularisation; he described understanding of science by decision-makers and citizens as 'a modern necessity' (Nielsen, 2019). Michelsen's Division existed only for five years but is taken as "part of a wider recognition at the time that science builds on communication and that science popularization is integral to science's communicative activities" (Nielsen, 2019, p.249).

In the United States, engineer and policy adviser Vannevar Bush offered a vision of science as the goose laying golden eggs, i.e., delivering economic wealth, social progress and military power if appropriately fed. In this context, popularisation was expected to 'sell science' to the broader public to strengthen social support and legitimation as well as to justify taxpayers' investments (Lewenstein, 2008). This fuelled the development of new popularisation strategies and channels, including interactive science centres and partnerships between science institutions, Hollywood and animation studios. At the same time, the diffusion of mass media, particularly television, offered new opportunities to communicate science to broad, international audiences. The general tone of science communication, in this age, was celebratory and optimistic: the achievements and promises of science were overall presented as a positive and progressive force. Walt Disney Studios produced cartoons in the 1950s, often in collaboration with state departments and agencies, for American TV channels about nuclear energy and space exploration. This 'Big Science' required massive state funding and at least public acquiescence, if not active support, so it gave added impetus to the emerging community of publicly communicating scientists.

During the 20th century, successful popular science books were authored by leading scientists and popularisers, including Nobel laureates like Alexis Carrell, Jacques Monod and Steven Weinberg, and highly acclaimed popularisers with strong scientific backgrounds like Carl Sagan and Isaac Asimov. In 1977, Sagan was invited to give the Royal Institution Christmas lectures, which were the platform for pioneering popularisers of earlier generations. As a photogenic, public-facing space specialist when space exploration was (literally) taking off, Sagan was the right person in the right place at the right time and soon became an international household name with his television series, Cosmos, broadcast in the United States from 1980 and subsequently in over sixty other countries. Sagan and television appeared made for each other, but his contributions to public communication were not just celebrations of science; like some other notable popularisers mentioned here, he also took time to reflect on the implications of doing this work alongside his research and university teaching, raising questions that remain pertinent to practice and education in science communication today. He advised his colleagues on humility: "Remember how it was before you yourself grasped whatever it is you're explaining. Remember the misunderstandings you almost fell into, and note them explicitly" (Sagan, 1996).

British immunologist Peter Medawar, who won the Nobel Prize for Physiology in 1960, maintained a highly visible public presence through popular science books, articles in newspapers and magazines and frequent radio and television appearances. His science commentary included science criticism as applied to the practice and organisation of science. In *Advice to a Young Scientist*, Medawar (1979) addressed the notions of superiority and genius associated with being a scientist, insisting on the importance of common sense and an enquiring mind.

The strengthening association of science popularisation and science critique among practitioners was matched with a growing critical awareness of the social and political implications of science, particularly for health and environment within public opinion. This shift found expression – and reinforcement – in communication about issues related to science and technologies (e.g. nuclear energy, use of chemicals in agriculture) by new actors and within new contexts: social movements, popular books written by scientists outside of academia. The case of marine biologist Rachel Carson is often regarded as emblematic: her book *Silent Spring* (1962) is represented as a key influence on the development of public awareness of environmental issues, often raised by the application of science. Another interesting case is the work of chemist and inventor James Lovelock: his Gaia theory, presented in technical publications and popular books, attracted the attention of interdisciplinary, cross-cutting audiences between science, culture and society.

Towards the end of the 20th century, science communication was significantly reframed in terms of institutional policy. In Europe, this reframing is often associated with the so-called "public understanding of science" era. Influential members of the scientific community sent out a call to arms. In their view, the general public was not attentive enough to science. Indicators of 'science literacy', measurements of basic knowledge of science content and methods, were often used and referred to in that respect. Citizens' lack of awareness and low literacy could result in a number of problems, including irrational suspicion, concern and outright opposition to new results and applications of science. A substantial communicative effort was therefore needed to improve public awareness and counteract such problems. Several national and international non-governmental organisations and governmental bodies created programmes dedicated to promoting and funding science communication initiatives such as science festivals, science cafés, and activities in science museums and centres.

In Chapter 4, we pick up the threads of science communication's development in the late 20th century as it came to be known by that name.

REFERENCES

Bucchi, M., 1997. "The Public Science of Louis Pasteur: The Experiment on Anthrax Vaccine in the Popular Press of the Time". *History and Philosophy of the Life Sciences* 19 (2), 181–209.

Butterfield, H., 1957. *The Origins of Modern Science, 1300-1800.* London: MacMillan.

Carson, R., 1962. *Silent Spring.* Boston: Houghton Mifflin Company.

da Rocha Gonçalves, D., 2020. "Science between the Fairground and the Academy: The Case of Dutch Science Popularizer L. K. Maju (1823-1886)". *Public Understanding of Science* 29 (8), 881–891. https://doi.org/10.1177/0963662520965093

Finlay, S. M., Raman, S., Rasekoala, E., Mignan, V., Dawson, E., Neeley, L., Orthia, L. A., 2021. "From the Margins to the Mainstream: Deconstructing Science Communication as a White, Western Paradigm". *JCOM* 20 (01), C02. https://doi.org/10.22323/2.20010302

Finnegan, D. A., 2017. "Finding a Scientific Voice: Performing Science, Space and Speech in the Nineteenth Century". *Transactions of the Institute of British Geographers* 42 (2), 192–205. https://doi.org/10.1111/tran.12159

Gascoigne, T., Schiele, B., Leach, J., Riedlinger, M., Lewenstein, B.V., Massarani, L., Broks, P., (Eds.), 2020. *Communicating Science: A Global Perspective.* Acton: The Australian National University Press.

Goldsmith, M., 1986. *Science Critic: A Critical Analysis of the Popular Presentation of Science.* London: Routledge & Kegan Paul.

Haldane, J. B. S., 1985/1941. How to Write a Popular Scientific Article. In J. B. S. Haldane (Ed.). *On Being the Right Size and Other Essays.* Oxford and New York: Oxford University Press.

Hamblyn, R., 2001. *The Invention of Clouds: How an Amateur Meteorologist Forged the Language of the Skies.* London: Picador.

Hudson, K., 1988. *Museums of Influence.* Cambridge: Cambridge University Press.

Huggins, M., 1907. Agnes Mary Clerke and Ellen Mary Clerke: An appreciation. Printed for private circulation. https://www.gutenberg.org/files/64057/64057-h/64057-h.htm

Jackson, R., 2018. *The Ascent of John Tyndall: Victorian Scientist, Mountaineer, and Public Intellectual* Oxford: Oxford University Press.

Kenyon, T. K., 2009. Science and Celebrity: Humphry Davy's Rising Star. Science History Institute, 23 December. https://www.sciencehistory.org/stories/magazine/science-and-celebrity-humphry-davys-rising-star/

Kirby, D., 2003. "Scientists on the Set: Science Consultants and Communication of Science in Visual Fiction". *Public Understanding of Science* 12, 261–278.

Lewenstein, B., 2008. Popularization. In J. L. Heilbron (Ed.), *The Oxford Companion to the History of Modern Science*. Oxford: Oxford University Press, 667–668.

Mantegazza P., 1881. La scienza nell'Italia Nuova. *In Annuario del R. Istituto di Studi superiori pratici e di perfezionamento*. Firenze: Le Monnier, 1–21.

Medawar, P. B., 1979. *Advice to a Young Scientist*. New York: Harper Collins Publishers.

Murphy, S., 2021. *The First National Museum': Dublin's Natural History Museum in the Mid-Nineteenth Century*. Cork: Cork University Press.

Nature, 1887. "Professor Tyndall and the Scientific Movement". *Nature* 36, 217–218. https://doi.org/10.1038/036217a0

Nielsen, K. H., 2019. "1947–1952: UNESCO's Division for Science & Its Popularization". *Public Understanding of Science* 28 (2), 246–251. https://doi.org/10.1177/0963662518787572

Orthia, L. A., 2020. "Strategies for Including Communication of Non-Western and Indigenous Knowledges in Science Communication Histories". *JCOM* 19 (02), A02. https://doi.org/10.22323/2.19020202

Poskett, J., 2022. *Horizons: A Global History of Science*. New York: Viking.

Raichvarg, D., Jacques, J., 1991. *Savants et ignorants, une histoire de la vulgarisation des sciences*. Paris: Editions du Seuil.

Sagan, C., 1996. No Such Thing as a Dumb Question. In C. Sagan, A. Druyan (Eds.), *The Demon-Haunted World: Science as a Candle in the Dark*. New York: Random House, 300–317.

Schiele, B., 2021. Science Museums and Centres: Evolution and Contemporary Trends. In M. Bucchi, B. Trench (Eds.), *Routledge Handbook of Public Communication of Science and Technology*. London: Routledge, 53–76.

Shapin, S., 1990. Science and the Public. In R. C. Olby, G. N. Cantor, J. R. R. Christie, M. J. S. Hodge (Eds.), *Companion to the History of Modern Science*. London: Routledge, 990–1007.

Turbil, C., 2017. "Paolo Mantegazza and the Dream of "making" Science Popular Circa 1860–1900". *Public Understanding of Science* 26 (5), 627–631. https://doi.org/10.1177/0963662517695117

Wells, H. G., 1894. Popularising Science. *Nature*, 50 (1291), 300–301.

3

SCIENCE COMMUNICATION
POLICY AND THEORY
DEFICIT AS DEFAULT

Science Communication discussed with my father

Hi son. I just dropped by to say hello, but I can see you're working. What are you up to?

Well, you know, I have been the editor of a journal, 'Public Understanding of Science'. It's been 3 years – I've told you quite a few times now.

'Public Understanding of Science'? Good for you. There's so much need for public understanding of science, these days. Climate change denial, vaccine scepticism … I keep reading about … what do you call it? Fake news. Actually, it seems to me you should call your journal 'Public Misunderstanding of Science' these days, don't you think? People are sceptical about vaccinations, GMOs, climate change, nuclear energy … anything that science proposes or suggests, you name it, they oppose it!

Hold on, dad. You are mixing up different questions here. Let's take nuclear energy as an example. Citizens may have the highest consideration of nuclear physicists and engineers and of their knowledge. At the same time, they may think that nuclear power is not the best option for their country for a number of reasons: because they are worried about long-term costs of waste disposal, about potential geological risks, about political and institutional instability. Do you agree?

Um … yes, I suppose that's possible. But what about vaccinations? People prefer to expose their children to enormous risks rather than believing the basic science facts. How do you explain that? This must be because they are ignorant. How can the individual decide about this sort of thing? We're not

DOI: 10.4324/9781032646749-3

talking about food preferences are we? Facts are facts, and Science is the same at every latitude!

Dad, are you a 'Deficit model' believer?

Excuse me? I might have some memory lapses now and then, but my brain still works perfectly well.

I am not talking about your own deficits. I am talking about the Deficit model. In our field, we call Deficit model the idea that people are sceptical about certain implications – or even applications – of science because they are ignorant about the science.

Well, this deficit, or whatever you call it, sounds quite reasonable to me. 'I know what I like, and I like what I know' ... was that the March Hare from Alice in Wonderland? (Excerpt from Bucchi, 2019)

"People are ignorant and distrustful about science"."If people would know more about science, or about a specific science topic, they would be more positive and trustful, open to new applications and implications of research and its technological outputs". "People are easy prey of misinformation on social media, including science related topics". Surely you have encountered these or similar assertions in the news media or through your social media feed; quite likely you have heard scientists or politicians make similar claims, for example, when expressing their concerns about citizens unwilling to vaccinate against COVID-19, or being concerned about new investments in nuclear power production. These claims are largely prejudices and stereotypes for which we find no substance in what we know about science communication and public perception of science. However, before we expand on this critique, we first need to outline the development of policy and theoretical approaches to science communication in order to better understand them.

Substantial reflection and legitimation of efforts to promote science communication became evident, particularly in the last decades of the 20th century. A number of factors conjoined to make scientists and their organisations and policy institutions aware of the need for more communication of science addressed to the general public. A relative slowing down of public investments in research after decades of sustained growth; and perception of increasing citizens' concerns for the environment as well as for potential negative consequences of technology are among the factors which can be identified to have contributed to such awareness.

"We need more science communication!" was the cry. In the anglophone context, the report on the Public Understanding of Science by The Royal Society (1985) is usually remembered as a landmark document. Scientists, organisations and institutions used several arguments to justify the need for more efforts and investments in public communication of science. These arguments were political, economic and practical. For example:

- Better-informed citizens make better decisions
- More communication will increase public support to secure public funding for research
- More communication will help recruit more university students to Science and Technology subjects to sustain the skills pipeline
- Better understanding of science is vital to democratic and cultural participation

Another reason for seeking to direct science communication towards such ends lay in the interpretation by those actors of the current state of science communication if left to its own devices. This was judged unsatisfactory, and two of the three parts involved (science, the media and the public) were mostly blamed: 1) the media, for being unable and unwilling to perform accurate delivery of scientific ideas and results from the scientific community to the general public; 2) the general public, considered to be uninformed and, as a consequence, sceptical, if not openly distrustful, of science.

The motivations and the stereotypes already described support definitions of science communication that stress institutional goals, strategic objectives, policy and political reasons. The purposes of science communication are commonly expressed in terms of translation from one sphere to another, of bridging gaps or breaking down boundaries between communities, of developing interfaces. It was on the basis of such rationale that, particularly since the 1990s, programmes were launched in many countries and internationally, sponsored by public (e.g. governments, the European Commission) and private organisations (e.g. foundations, scientific academies) to promote science communication and increased public awareness and understanding of science. The wide range of activities included specific communication initiatives, training courses for journalists and scientists, media resource services assisting journalists in finding appropriate expert sources, conferences and roundtables. Soon, and interestingly, the original justifications were forgotten. Science communication became an end in itself; more science communication became an imperative, worthy of collective efforts and investments.

THE DEFICIT MODEL AND ITS DISCONTENTS

The "science communication wave", which surged in the late 20th century, was largely based on assumptions about communication, the media, audiences and society at large, as well as science and science communication itself, that have often been referred to, more or less simplistically, as the "deficit model". To better understand such assumptions, we should break them down more clearly, as they often mix and merge different aspects of what we prefer to call a *diffusionist conception of public communication of science*. This notion incorporates several different notions:

- the media as a channel designed to convey scientific notions but often unable to perform this task satisfactorily due to lack of competences and/or predominance of other priorities (e.g. commercial interests);
- the public as passive, whose default ignorance and hostility to science can be counteracted by appropriate injection of science communication;
- science communication as a linear, one-way process in which the source context (specialist elaboration) and target context (popular discourse) can be sharply separated, only the former influencing the latter;
- communication as a broader process concerned with the transfer of knowledge from one subject or group of subjects to another;
- knowledge as being transferable without significant alterations from one context to another, so that it is possible to take an idea or result from the scientific community and bring it to the general public.

Such notions have mutually reinforced each other and have sometimes overlapped to some extent, but one of the labels most frequently used to refer to this constellation of notions, the "*deficit*" model, refers in particular to the second assumption above. All five, however, have been extensively challenged during the past three decades by empirical research and theoretical reflection. Some of these critiques highlighted that the deterministic assumption "more information=more support of science fields and their implications" does not hold: one can be trained in physics, for example, and still be sceptical of nuclear energy as a viable solution in a certain political or geophysical context. Indeed, the opposite can often be true: more information can be associated with more critical concerns. Other critics argued that the prevailing practices of popularisation and dissemination contained an assumption of deficit on the public's part and that this was inhibiting and inappropriate. This critique was fundamentally moral or philosophical, rooted in commitment to equity and accessibility.

Already in the early 1980s, before the public understanding of science movement had adopted this name, the direction of such efforts was being questioned and contested. Science writer Leon Trachtman (1981) described the emerging movement as engaged in "missionary activity" based on the belief that people will make more intelligent consumer decisions and participate more effectively in democracy if they are better informed about science. He noted that scientific knowledge is often unclear or ambiguous, and that the public understanding of science effort, as he called it, misrepresented this and other aspects of science. Drawing strongly on Trachtman, communication scholar Christopher Dornan (1990) critiqued the transmission and translation approach to media science, suggesting that the increased science communication project was allied with "the efforts of scientific organisations to engineer dutiful coverage" (p. 62) of science, based on deference to it; Dornan referred to this as "an ideological labour" (p. 63).

Sociologist Stephen Hilgartner (1990) described the dominant view of science popularisation as rooted in a notion of pure scientific knowledge that was degraded or distorted through simplification for popular audiences. This view, he argued, is itself an oversimplification of the process through which scientific knowledge spreads, including among scientists. Physicist Jean-Marc Lévy-Leblond (1992) also referred to communication among scientists as part of the bigger picture of popularisation, insisting there was "no single general knowledge gap between scientists and non-scientists, but there is instead a multitude of specific gaps between specialists and non-specialists in each field" (p. 17). He sought to "stress the necessity of more sophisticated and critical approaches" (p. 21).

However, despite the availability of these several examples of critical approaches, the dominant views and assumptions survived strongly through the following decades. The undoing of the deficit model, when it took root, left deference to science in place, albeit science available to dialogue, which was elevated to the replacement model. By 2000, the need for such a replacement was so widely accepted and amplified that the seminal House of Lords report of that year could refer, on the basis of its wide consultation with science communication communities, that there was "a mood for dialogue". Deficit was said to be outdated, superceded by dialogue and – a slightly younger cousin – public engagement with science. But still, deficit ways persisted, continuing to influence activities and funding programmes. In some cases, they were found behind the cloak of dialogue. In others, they continued as if they had never been challenged. In 2016, the journal, *Public Understanding of Science,* sought essays in answer to the question, "Why does the deficit model persist?". The phrasing of the question indicated clearly that the deficit model's time had passed, or at least should have passed. It also reflected how, in a certain historical phase, the deficit model has become the "evil" of science communication, a synonym of all sorts of its negative aspects, a sort of straw man to blame in order to move forward with communication strategies.

MISINFORMATION ON MISINFORMATION?

In the first decades of this century, thinking and practice in science communication had to take into account a new landscape which was characterised – among other things – by the advent and fast diffusion of digital technologies (particularly smartphones) and social media like Facebook, Twitter (now X), Instagram and TikTok. Also, in discussions about unexpected political results in 2016 like the US presidential elections and the UK pro-Brexit vote, concern mounted about the spread of misinformation (sometimes shortened and sloganised as "fake news") through social media and its potential impact on political opinions.

Once again, a series of assumptions and equations were quickly made: social media=misleading information; misleading information= biased opinions (and voting). Social media were increasingly blamed for hosting propagandistic posts, considered key factors to account for otherwise unexplainable outcomes such as the pro–Brexit vote. Similar assumptions were extended with regard to science communication and related issues: social media=misleading information about trust; misleading information=mistrust of science and scientists. Further stereotypical characterisations of the science/citizens communicative nexus were, and still are, often added to the picture:

general lack of trust in science, scientists and scientific results; dependence on social media for information, including science-related information.

These and other related concerns escalated worldwide during the global COVID-19 pandemic. An alleged "infodemic", particularly in social media and in combination with an alleged longstanding mistrust of science and scientists, was considered a gateway for resistance to restrictive protection measures and to vaccination against COVID-19. Again, these characterisations have been convincingly challenged by several studies, both in general and more specifically in relation to science communication. At the general level, such studies have pointed out, among other things, that the level and spread of misinformation (in itself, difficult to define) tends to be exaggerated and projected on other people rather than oneself: "the others are gullible" (e.g. Altay, Berriche and Acerbi, 2023; Altay and Acerbi, 2023). In the context of science communication and public perception of science and related topics (particularly health issues), people tend to be more critical of social media content credibility than normally assumed. Questioning again a widespread stereotype, citizens in many countries turned mainly towards traditional media sources and institutional sources for information about the pandemic and related protection measures (including vaccination) and only occasionally towards social media.

On the question of trust in science and scientists, surveys in Europe, the United States and elsewhere have shown that it has been consistently high and increasing during the past two decades, with science and scientists often trusted by more than 80 per cent of the population (e.g. Askvall et al., 2021; Observa, 2024). Despite the evidence of stable or increasing public trust in science across many and very diverse cultures, however, the talk of a crisis of trust or a climate of mistrust persists and the agents of distrust, e.g. in relation to climate science denial, get disproportionate attention. In the international science communication community, there is a rising interest in public trust in science as a policy and research issue, and this is also generally accompanied by a normative view of the crisis or deficiency of public trust in science and the role of science communication in boosting such trust. A leading national representative body in science, the American Association for the Advancement of Science, and an international representative body, the International Science Council, have based programmes and projects on tackling the public trust problem. Several recent and current European Union-funded projects were or are aiming to develop the means to boost public trust in science and to improve science's trustworthiness.

The discourse around public trust in science echoes diagnoses of declining or deficient trust in political and other institutions, such as frequently declared by Antonio Guterres, United Nations General Secretary, and claimed by PR consultancy Edelman, which promotes its Trust Barometer with talk of the "lack of faith in societal institutions triggered by economic anxiety, disinformation, mass-class divide, and a failure of leadership"[1]. The supposed trust-in-science deficit adds to the various other deficits that have been ascribed to the publics for science, including in literacy, awareness, understanding and information. The common thread through these topics is that society is deficient in relation to science, and the scientific community and its allies need to – and can – redress this deficit. The preoccupations with public misperceptions and misinformation and disinformation have further reinforced such "deficit" views. Wynne (2006) described the "public mistrust of science" problem as "repeatedly projecting the blame onto incompetent publics, irresponsible and misinforming media, and non-governmental organizations, as well as other convenient scapegoats".

This, of course, does not mean that there is not an important issue of quality and a corresponding challenge in contemporary science communication. But the challenge is not about policing science news or clinically debunking individual pieces of (mis)information – which would raise, among other things, the question of who has the right and the appropriate criteria to identify misinformation. Nor can it be solved through the simplistic (and fallacious) notion that traditional media is equal to good quality and social media to poor quality. Traditional news media are not immune from dubious information and content, just as social media can host high-quality lectures and other materials.

In terms of general information and opinion formation, the quality challenge arises in connection with substantial changes in citizens' habits and the collapse of the traditional new media business model. A striking contemporary paradox is that everyone complains about the (poor) quality of information, but few want to pay for it (for example, by buying a newspaper subscription or magazine). The quality challenge in contemporary science communication is broader than it is normally thought, with accuracy being just one of the many dimensions of quality. The challenge has to be read in the context of the "crisis of mediators", which is the result of a two-fold combination of forces. The first force is, clearly, the changing media landscape: digital platforms in their variety, disabling traditional forms of social action (e.g. buying a newspaper aligned with one's own political preferences; watching evening news) and enabling new ones (scrolling social media/contact feeds; following video or podcast channels); shifting power from media content producers/publishers to an oligopoly of tech businesses (Google, Meta, Tik Tok, Twitter/X). We return in chapter 7 to the quality question as one of several key challenges facing science communication. The second force, not less relevant, is the increasingly strong "third mission and public engagement" push in academic and research institutions. Again, studies have documented (e.g. Entradas and Bauer, 2022) the increasing amount of financial and staff resources devoted by these institutions to communication with non-experts. The traditional imperative to "publish or perish" is supplemented by "engage or perish".

These efforts and initiatives aimed at public audiences do not generally go through the traditional mediation of journalists and popularisers but are placed – either through digital media or physical events – "direct to consumers". From this follows the articulated and multidimensional challenge of quality for contemporary science communication. In the context of mediated science communication, quality was assessed through the reputation of the channel: if it's on the BBC, or in a science museum, or in *Frankfurter Allgemeine*, *Le Monde*, *Il Corriere della sera*, it has to be credible and worth attention. This shortcut is not available anymore. What do I get on the social media channels or web pages of a research institution: the state of the art of a certain science field or the self-promotion of their own activities?

The matter is even further complicated by the multiplying opportunities for communicative activism by individual scientists. Who are they speaking for? On behalf of their own institutions or just for themselves as independent scholars? And what happens when they comment on topics outside their specific speciality area (other research fields, or political and social issues at large)? The pandemic crisis has further amplified these dilemmas and ambiguities and put them under the spotlight. While trust in science, scientists and scientific institutions is high and increasing, a much more critical judgement of scientists' communicative performances often emerges, particularly with regard to increasing personalisation and potential controversies among experts. We will come back to this and related points when we consider the ranges of actors and platforms for science communication in Chapter 5.

OH, DEFICIT, DEFICIT, WHEREFORE ART THOU DEFICIT?

So, if it has been repeatedly challenged and officially superseded, why does the deficit approach always return, even if sometimes in disguise? Why do (unproven) stereotypes of the public and science communication persist? We consider that satisfactory explanations for how these ideas have become embedded in the cultures of scientific communities and thus in the communities of science communication can be found in such factors as the socialisation of scientists, the growth of modern science communication as an arm of science and ideologies of scientism. These are all factors in sustaining deficit thinking as a default perspective for scientists and their agents.

Scientists are socialised through their education and their professional communities, absorbing values and norms that buttress a view of science as special and apart. Science communication has grown in close proximity to scientific organisations and institutions and, even if recruitment to this field has gone wider, the prevailing values and norms of the scientific communities have affected it deeply. In these contexts, the deficit view is assumed as self-evidently valid, and in this sense, it is ideological, also in the sense that it implicitly supports privileged social status. Ideology appears to its holder as natural, or common sense. The holder may not recognise themselves in the description of a follower of an ideology, in this case, the ideology of scientism. For such a person, it is self-evident that science, and only science, is a reliable way of knowing the world and is by definition superior to any other forms of knowledge. The supposed adherence to The Scientific Method is taken as a badge of belonging to that system.

A view of science as deserving to be loved and defended is the starting point of much public communication of science. State programmes in science communication have been a key element in the growth and stabilisation of the field. Typically, such programmes seek to build support for science, belief and trust in science. In the spirit of these programmes those who question aspects of science may be cast as anti-science. A naiver form of scientism supporting deficit-like views can also be identified in the common assumption that science communication is "easy" and can be improvised, particularly by those who have a background in the "harder" sciences.

Deficit views and stereotypes of an (imaginary) anti-science public are also ideological in functional terms. They exempt, in many ways, actors and institutions from improving the quality of their own science communication – after all, if people are distrustful and incapable of understanding, why bother? In this sense, they become a "self-fulfilling prophecy", as theorised by the founder of the sociology of science, Robert K. Merton, resulting in low-quality science communication that reinforces such stereotypes. We have seen, unfortunately, multiple examples of this attitude during the pandemic: improvised, extemporaneous communication by scientists and institutions, without any intelligence or knowledge of audiences, often based on prescriptions and emotional appeals rather than careful explanation (e.g. "you should vaccinate because it's your duty"; or "you should vaccinate to the benefit of your older relatives").

In a broader sense and beyond science communication, ideologies of misinformed and gullible audiences provide justification for institutional communicational failures. If the Brexit vote can be attributed to misleading posts on social media and to easily misguided citizens, then institutions that have failed, for example, in communicating the advantages of being in the EU for the UK are easily exempted from self- and external criticism.

We should be careful, however, not to confuse the deficit as ideology with the deficit as communicative mode. In this second meaning, the deficit has indeed still a place, for example, when experts or their institutions begin popular communication on a new topic or theme which is not yet visible or sensitive, addressing a deficit of information among general audiences. As such, the deficit could "peacefully" coexist within the broad spectrum of communicative interactions about science.

Once again, viewing science communication more broadly as the social conversation about science softens the missionary urgence of "delivering the message of science" across audiences. Science communication is much more than that, including several configurations (science in popular culture) that are not purpose-oriented but still can powerfully shape, in non-linear ways, public perception and attitudes. We expand further in Chapter 7 on the range of science communication models and their applications and implications.

NOTE

1. From the introduction to the 2023 Europe Report, posted at https://www. edelman.be/insights/2023-trust-barometer-europe-report

REFERENCES

Altay, S., Acerbi, A., 2023. "People Believe Misinformation Is a Threat Because They Assume Others Are Gullible". *New Media & Society* 26 (11), 6440–6461. https://doi.org/10.1177/14614448231153379

Altay, S., Berriche, M., Acerbi, A., 2023. "Misinformation on Misinformation: Conceptual and Methodological Challenges". *Social Media + Society* 9 (1). https://doi.org/10.1177/20563051221150412

Askvall, C., Bucchi, M., Fähnrich, B., Trench, B., Weißkopf, M., 2021. Trust in Science: Assessing Pandemic Impacts in Four EU Countries. https://pcst.co/doc/TrustInScience.20210114.pdf.

Bucchi, M., 2019. "Science in Society Discussed With My Father: A Parting Editorial in the Form of a Dialogue". *Public Understanding of Science* 28 (5), 514–518. https://doi.org/10.1177/0963662519851612

Dornan, C., 1990. "Some Problems in Conceptualising the Issue of 'Science and the Media'". *Critical Studies in Media Communication* 7 (1), 48–71. https://doi.org/10.1080/15295039009360163

Entradas, M., Bauer, M. W., (Eds.), 2022. *Public Communication of Research Universities. 'Arms Race' for Visibility or Science Substance?* London: Routledge.

Hilgartner, S., 1990. "The Dominant View of Popularization: Conceptual Problems, Political Uses". *Social Studies of Science* 20 (3), 519–539. https://doi.org/10.1177/030631290020003006

Lévy-Leblond, J.-M., 1992. "About Misunderstandings About Misunderstandings". *Public Understanding of Science* 1 (1), 17–21. https://doi.org/10.1088/0963-6625/1/1/004

Observa, 2024. *Annuario Scienza Tecnologia e Società*. Bologna: Il Mulino.

The Royal Society, 1985. The Public Understanding of Science. https://royalsociety.org/-/media/policy/publications/1985/10700.pdf

Trachtman, L. E., 1981. "The Public Understanding of Science Effort: A Critique". *Science, Technology, & Human Values* 6 (36), 10–15. http://www.jstor.org/stable/689093

Wynne, B., 2006. "Public Engagement as a Means of Restoring Public Trust in Science—Hitting the Notes, But Missing the Music?". *Community Genetics* 9 (3), 211–220. https://doi.org/10.1159/000092659

4

HOW SCIENCE COMMUNICATION BECAME A THING

In the early months of the Covid-19 pandemic Jacinda Ardern, prime minister of New Zealand Aoeteroa, became one of the first political leaders to acknowledge explicitly the need for science communication and communicators: "Without our science communicators to publicly inform, explain, teach, decode, counter misinformation and debate science matters many would remain in a space where they don't have [the] information they need, leading to poor choices being made at really crucial times" (Ardern quoted in LeBard, 2020). With this recognition science communication achieved a new visibility.

In this chapter, we examine the emergence of science communication as a social phenomenon with distinct organisational structures and community ways of thinking and speaking. Through the parallel growth of voluntary activities, professional practice, formal education and research in the field, "science communication" came to be a recognised name for sets of evolving activities and issues. In considering science communication becoming A Thing we are adopting a common current usage in English that is occasionally ironic, signalled by accompanying "air quotes", to refer to a phenomenon that has achieved significant social recognition. Becoming a thing can happen to an idea, like workplace diversity, a culinary custom, like avocado on toast for breakfast, or cultural manifestation, like social media memes, or, indeed, anything one can imagine as having social resonance. Ceasing to be a thing is also possible, as US journalist Alexandra Kitty (2018) claims of her own once "noble profession" which was "an important thing, a necessary thing, a thriving thing" but, as she sees it, has lost all credibility.

DOI: 10.4324/9781032646749-4

The naming of science communication in the academic literature about science-society relations was a reflection, possibly also a driver, of its social emergence. In our edited anthology of foundational essays, papers and chapters on public communication of science (Bucchi and Trench, 2016) we charted the trend in the naming of the topic. The total 79 items in the collection were divided nearly equally into those published in 1994 or earlier (37) and those published in 1995 or after (42). "Science communication" was named as such in two of the entries in the earlier phase, but in 19 of those from the later phase; 17 of those 19 references came in entries from 2001 or later.

As it happens, the chosen threshold year, 1995, was when an academic journal titled *Knowledge* was renamed *Science Communication*. Three years earlier, a journal had been established in Britain to support the growing community of scholars interested in communication within a science-society context; it was – and remains – titled Public Understanding of Science, reflecting the prevailing perspective on communication issues at that time.

In wider social spheres science communication's arrival was founded on a number of roughly simultaneous developments such as the establishment of government, state agency and scientific society programmes to boost public awareness of science; the building of interactive science centres; the provision of training courses and accompanying manuals for publicly-communicating scientists; promotion of media attention to science; the establishment of higher education courses for education of science communication professionals. These and related steps being taken in many countries were taken as indications of the "global spread of science communication" (Trench and Bucchi, 2021). A large-scale survey of modern science communication across 39 countries (Gascoigne and Schiele, 2020) included further criteria such as the formation of science communication associations, PhD research in the field and employment opportunities for science communicators. The verifiable presence of initiatives such as those listed is the mark of science communication becoming a thing in diverse communities, cities and countries.

The internationalisation of science communication was reflected in the almost simultaneous adoption across many countries of the term itself as *communication scientifique, Wissenschaftskommunikation* and their equivalents in other languages. It is important to note, however, that other terms continue to be used – in some cases more than science communication – such as the social appropriation of science, scientific culture and science popularisation. Of this variety, Gascoigne and Schiele (2020, p. 12) observe that "the heterogeneity of practices implies the heterogeneity of the field, and it is this heterogeneity that has led to the wide variety of terms being used for 'science communication'".

Whatever the terminology, we can state that science communication has become a thing on a global scale. The key contributors to this process include governments, state agencies and business interests seeking to attract enrolments in science and technology courses (increasingly referred to as STEM, to represent science, technology, engineering and mathematics) and thus boost the supply of suitably qualified people for high-tech industries; higher education institutions responding to this drive; scientific societies aiming to build support for scientific endeavour and thus also for their claims on increased public resources; philanthropic foundations, associations of scientists and of science-interested people committed to spreading the word about science's achievements. The aggregated efforts of these various actors across several sectors have contributed to the growth and diversification of science communication practices and structures.

SCIENCE COMMUNICATION EDUCATION PROGRAMMES

In becoming a thing, science communication is a recognised element of job specifications, it offers career paths and professional mobility, and it is a focus of consultancy business. In the related contexts of employment and education, a useful marker of science communication's emergence is the establishment and evolution of university courses in the topic. The spread of such courses, their variations in character, their recruitment of students and the subsequent paths of their graduates are all useful indicators of science communication's status. The different experiences of two such university courses in which the present authors have or had leading roles may be instructive in this regard.

The Masters in Science Communication at Dublin City University (DCU), Ireland, enrolled its first students in 1996 at a time when such programmes were counted in single figures around Europe, and in single figures outside that continent. The DCU Masters was established in partnership with Queens University Belfast (QUB), Northern Ireland; the two coordinators were a particle physicist at

QUB and a journalism lecturer at DCU. The initiative was driven by the personal interests of these two individuals but also reflected the growing attention in both jurisdictions to public awareness of science. It attracted students from various countries and disciplinary backgrounds, though mainly – and increasingly – from natural sciences; graduates in these subjects were looking for career opportunities in, or related to, science but outside the laboratory.

Early contributors to the teaching included a science historian, a psychologist, an archaeologist, a physicist (at QUB), a biotechnologist, a chemist, communication and cultural studies scholars and journalism educators (at DCU). This mix was facilitated mainly by individual interest and goodwill, including from senior management, though the inter-institutional and interdisciplinary cooperation were not formally endorsed. The curriculum has always included elements of communication practice and media analysis but was also always more than a training suite to convert science graduates into science communicators.

As with other such programmes across the world, graduates have found outlets for their interests and aptitudes in science centres, various branches of mass and specialist media, technical writing and public information and engagement roles with scientific societies and research centres. Gradually, employers in these sectors were coming to acknowledge a qualification in science communication as either desirable or essential for positions on offer. It is also worth noting that a small but consistent minority of graduates returned to the practice of science but with an emphasis on public communication, and a similar proportion went into training, teaching or research in science communication. The two latest coordinators of the DCU programme, as of 2024, were both graduates of it, who went on to complete PhDs in science communication topics.

While the DCU Masters is based in a School of Communications, the Masters in Communication of Science and Innovation at the University of Trento, Italy, has its home in the Department of Sociology and Social Research. This programme enrolled its first students in 2018 and quickly had to adapt to pandemic circumstances. It is now delivered exclusively through distance learning methods and in English, which has enabled it to operate as an international programme with students from all continents. Different from science communication Masters established in previous decades, this course does not aim at training science journalists or science communicators to work in media outlets. Due to the changes described in the previous chapter, such working opportunities have significantly decreased. On the other hand, there is a growing need for science communication competencies within research and academic institutions to plan engagement and outreach strategies and help scientists disseminate their results. To this end, students are trained to be knowledgeable about three main areas: media and their operational logic; research organisation; audiences and society.

These two programmes illustrate the similarities and differences between the Masters around the world that have become key markers of an established science communication infrastructure in their respective countries. However, they also illustrate that such programmes are more than national; in many cases, they draw students from across their continent, or even from several continents. Science communication, in becoming a thing, has become international and intercultural. The English language is the vehicle of this internationalisation: professional education in many non-anglophone countries is offered in English as a means to accommodate international students.

The two programmes outlined above also highlight the question, what does it mean to learn this thing called science communication? Studies of nearly two decades ago aimed to compare the then-existing Masters on criteria of preparing students for specific tasks and duties in their future employment. But for some programmes, at least, the emphasis has been equally placed on deepening students' (critical) understanding of science's place in culture. A European Union-funded project of the early 2000s, ENSCOT, developed teaching materials that spanned these two broad objectives and EU-funded projects of the mid-2020s, Globalscape and Coalesce, have collected detailed information about the many and diverse professional education offerings around the world, and sought to chart their possible future directions.

A set of papers on science communication education arising partially from the Globalscape project examined issues in this sector (JCOM, vol, 22, issue 06), including integrating sustainability into the curriculum, the geographic distribution of such education programmes, graduates' perceptions of the value of the acquired knowledge and skills to their employment. One of the present authors (Trench, 2023) examined the place accorded within science communication education to interdisciplinarity, both as a way of working and as a topic for analysis. This essay, presented in terms of the "Premise, Promise, Problems and Pleasures" of interdisciplinarity made a "Proposition" as to how interdisciplinarity can be given greater prominence in higher education science communication programmes.

ROOTS OF SCIENCE COMMUNICATION STUDIES

The expansion of science communication postgraduate education has been underpinned by the growth in science communication research. Early contributions to the formation of this field from the 1960s onwards included those from sociology, philosophy and psychology. French social psychologist Serge Moscovici studied public perception of scientific ideas – initially, psychoanalysis – through the lens of "social representation" (1961). His further elaborations of this conceptualisation of social groups' shaping of ideas from intellectual domains directly influenced later work on public attitudes to science (see e.g. Bauer and Gaskell, 1999). American physicist and philosopher of science Thomas Kuhn's study of change and revolutions in the history of science (1962) showed the role of communication among specialists and in society more broadly in supporting and shifting paradigms in different scientific fields. German social theorist Jürgen Habermas, as he worked through his analysis of the transformation of the bourgeois public sphere from the 1960s onwards, reflected on the changing relations between experts, politicians and the public, noting that the sciences "have made a basically unsolved problem out of the appropriate translation of technical information even between individual disciplines, let alone between the sciences and the public at large" (1971, p. 69).

Modern research, Habermas wrote, is affected by "bureaucratic encapsulation" that has eliminated "the unconstrained and formerly automatic contact of the individual scientist with the larger public" (p. 76). He speculated that physicists might use *Time* magazine to keep abreast of developments in technology and chemistry and argued that it was important to keep communication open between disciplines "even if information from one specialist to another has to take the long route of ordinary language and the everyday understanding of the layman" (p. 77). Habermas thus offered early insight into the inextricable connection between the challenges of inter-scientific and extra-scientific communication. In the years after that, these two communication contexts came to be seen as separate – hence the emphasis on public communication of science (and technology) as distinct – only for more recent work to insist again on their interconnectedness.

The pioneering work in the 1930s of Polish microbiologist and philosopher of science, Ludwik Fleck, in the 1930s was rediscovered a half-century later. Fleck analysed the different layers of communication and communities around the development of a scientific fact. Different "thought collectives" – communities that share a certain "thinking style" – intersect repeatedly. An esoteric circle (that of specialists) and an exoteric circle of non-specialists gravitate around a certain style of thought that draws strength precisely from the continuous exchange between these circles. Among astrophysicists there can be doubts and distinctions, ambiguous observations and data: today for the general public the "Big Bang" represents the origin of the universe, full stop. The researcher, as a simultaneous member of different collectives of thought (the community of specialists to which he belongs, but also a party, a social class, and a certain cultural milieu), is at the centre of these continuous exchanges. "A risky experiment must be explained by the doctor in simple language in the way one person converses with another," Fleck wrote (1935/1979). His work connects with ideas later explored in science and technology studies (STS) about the hybrid networks and discursive devices involved in the development and establishment of scientific facts.

More focused studies of scientific literacy, visible scientists and science public relations from the 1970s and 1980s all fed into the emerging field of science communication research, along with critical perspectives on the focus on public understanding of science as being coupled with misunderstandings of the public.

PUBLIC LITERACY IN SCIENCE

A central concern of science policy particularly in developed economies and of the emerging science communication movement from the 1970s was the scientific literacy of the population. This was the focus of public opinion survey research conducted principally by state agencies. In the US, Jon D. Miller carried out and analysed numerous such surveys refining the questions beyond testing for knowledge of presumed basic scientific facts and diving deep into the demographic differentiation in the results. Even as he allowed for different understandings of the concept itself, his starting point, as for many others, was that public scientific literacy was low, even "deplorably low" and among a large section "extraordinarily low" (Miller, 1983). This, he argued, had implications for democracy and, having divided the population into a pyramid of decision-makers at the top down through policy leaders and an attentive public to the 80 per cent that is the nonattentive public, he urged that priority be given to expanding scientific literacy among the attentive public.

Over the past forty years there have been intense and, at times, testy debates on the value of survey questions about knowledge of scientific facts and theories and on the appropriate questions to include on attitudes to science, scientific institutions and scientists. An early and enduring contribution to that debate came from John Durant, one of, if not the, first designated professor of science communication, later a museum director. To the question, What is Scientific Literacy?, Durant (1994) answered that it included "knowing a lot of science" – the most familiar approach, he added – "knowing how science works" focusing on the processes, and "knowing how science really works" looking at science as social practice. Durant emphasised the role of the last of these as a corrective to "idealistic images" of the scientific attitude and scientific method. Thirty years later, such idealistic images still prevail and science communication has to face up to its role in their lasting influence.

Whereas scientific literacy studies commonly gave their main attention to the illiteracy or ignorance of the public, some sociologists of science focused attention on what sections of the public do know. As demands rose for greater public recognition of scientific expertise, Wynne (1992) showed through a case study of northern English sheep farmers' responses to scientific advice about restrictions arising from the 1986 Chernobyl nuclear power station accident that formal scientific expertise is not always credible to its audience and supposedly lay (as distinct from clerical) people have their own local knowledge and can reflect on their relationship to institutions of authority. Wynne argued that this requires reform of scientific institutions to, inter alia, recognise "socially extended peer groups legitimated to offer criticism of scientific bodies of knowledge".

In a study of AIDS activism in the United States Epstein (1995) described such criticism – in this case, from patient and advocacy groups primarily in gay communities – as "the construction of lay expertise". The contribution of these activists was most keenly felt in the reform of clinical trials of AIDS treatments. Anthony Fauci, then director of the National Institute of Allergy and infectious Diseases and much later the central figure in the US authorities' handling of the COVID-19 pandemic, conceded that some of the activists "are brilliant, and even more so than some of the scientists" (personal interview, quoted in Epstein, 1995). Epstein's analysis and related studies of public controversies around science- and technology-related issues challenged science communication orthodoxy on who gets to speak about science and who is expected to just listen.

MEDIA ATTENTION TO SCIENCE

Sociologists Allan Mazur and Dorothy Nelkin were among those attending to such controversies in the 1970s and 1980s but Nelkin shifted focus to communication-focused studies, including in her critical account (1987, 1995) of how science journalism "sells science", retailing rather than investigating, as she put it. She examined the use of language, particularly metaphors, in media reports of science and then (1995) turned the spotlight on the "promotional metaphors" deployed by scientists in public communication of genetics. Some of their rhetoric was, she wrote, "overblown" and some metaphors "misleading".

Alongside critical studies of media and public relations content like Nelkin's, however, the dominant trend in studies of media science remained a lament about the low levels of media attention to science, a story retold many times as more countries came into the frame. Rather as with claims of low scientific literacy, these claims of low media attention tended to lack the context to allow "low" or, for that matter, "high" to be understood. These claims relate to the deficits of public understanding and trust we explored in Chapter 3.

The concept of visible scientists was introduced by Rae Goodell (1977) to draw attention to a new media presence of scientists. They are those members of the scientific community – including, for example, some Nobel Prize winners – who have great recognition and influence not only among their colleagues but also in the media and the general public. Achieving this visibility required an ability to satisfy some operational logics of the mass media, to speak in public with charisma, and to respond quickly to media requests. Since then, this visibility has undergone significant transformations in terms of intensity and methods of public presence, also in relation to the possibilities offered by means of communication such as social media. The pandemic crisis has amplified and made this trend even more macroscopic, accentuating the visibility of some researchers and bringing others into the spotlight, highlighting and accentuating forms of personalisation of communication.

Two aspects of the personalisation of communication, and their consequences, deserve particular attention. Firstly, the public visibility of scientists presents, in an even more accentuated form, the dynamics of concentration and accumulation identified by the founder of the sociology of science Robert K. Merton within scientific communities. Merton (1968) characterised these dynamics as the "St. Matthew effect", from the passage of the Gospel of the same name in which it is said: "To him who has, it will be given, and he will have abundance: but from him who does not have, even what he has will be taken away" (Matthew, 25, 29). Those already in positions of visibility and recognition will therefore have privileged access to other resources and positions of visibility, and so on. A scientific contribution will have greater visibility if presented by an already well-known scientist compared to colleagues who have not yet made their mark. In the words of a Nobel Prize winner in physics quoted by Merton, "the world tends to give credit to people who are already famous."

Today the Matthew effect is accentuated and amplified through social media and in the increasing short circuits between the research sector and the communication sector. Science itself is thus, at least in part, permeated by a logic not dissimilar to that increasingly characteristic of the world of entertainment or sport: male and female scientists who have become familiar to the general public are often invited to comment on topics far from their expertise specifications.

The second aspect is the fragile balance between individual communication and visibility and communication and visibility of the institution to which one belongs. On whose behalf is the scientist speaking on a television talk show, interviewed by a news programme or speaking on social media or personally, in her research sector, or in the institution to which she belongs? These issues become even more delicate when they deal – as during the pandemic crisis – with issues that are relevant to citizens, or sensitive on a cultural, social and political level. Hence numerous debates and even proposals, which are hardly realistic on a legal level, have arisen on limiting the communicative presence of researchers or subordinating it to the authorisation of the institution they belong to. Also of note are the possible consequences that this visibility – in its various forms – can have on public opinion in terms of public perception of science and trust in science, scientists and research institutions (Bucchi, Fattorini and Saracino, 2022).

Research themes of the 1990s in science and technology studies, such as citizen science (e.g. Irwin, 1995), contemporary science as Mode-2 (Gibbons et al., 1994) or post-normal (Funtowicz and

Ravetz, 1993), and new social contexts for science such as open science and civic science have all provided intellectual stimulation for the development of science communication as a field of education and research. A common thread in much of this work is the relocation of science within society and a departure from ideas of the two as separate domains, somehow connected by communication.

DIVERSIFICATION AND DEFINITION

The limits and harms of popularisation and deficit models, as discussed in Chapter 3, were part of the conceptual foundation of science communication studies. The naming of the deficit model and the discussion of other models, such as contextual, dialogical and participatory, structured the developing field and the intellectual community growing within it. From the early 2000s research was being done in the name of, and focused on, science communication that looked at the diversification of perspectives, practices, formats, models, media and publics. The rapid development of internet platforms such as email, bulletin boards and other forums, websites, online repositories and, perhaps most critically, social media required reframing of many of the received ideas and relationships.

The global spread of science communication was driven by shared ideas and aspirations related to the knowledge society and sustainable development. But this has also brought attention to issues of cultural diversity, which have also been made salient by more recent societal concerns of diversity, equity, and inclusivity (DEI).

These developments have not been linear or uniform. They do not represent a diffusionist process of globalisation as this has generally been understood in business and politics. The DEI concerns have prompted challenges – not only in science communication – to received ideas about the "scientific revolution" in Europe of the 17th century and the "modern science" which developed from it. The increasing acknowledgement of colonial exploitation of lands and cultures is accompanied by recognition of knowledge systems that preceded and were largely displaced by western modes.

All of this adds intensity to the question of what is the science in science communication, it having been widely assumed to be mainly the physical and natural sciences as they evolved in more developed economies. The social sciences and humanities have rarely been given focused attention in this context, much less the knowledge deriving from cultures that predate those shaped by modern science.

In reflecting on itself and its limits science communication is moving towards an adult stage, a grown-up thing with its own institutions, professional mobility, international networks, student and teacher exchanges, a growing library of studies and essays, and becoming more complicated.

One of the recurrent questions in science communication's becoming a thing has been: in the intellectual and academic context, is it a discipline? There have been diverse and shifting answers to that question often using "field" as the generic term, representing a reality that is less strictly delineated than a discipline. Logan (2001) considered science communication a sub-discipline of mass communication which, with that broader field, "represent emerging, pioneer, and somewhat unsettled social sciences"; Hornig Priest (2010, p. 3) underlined science communication's hybrid status as "inherently both multi-disciplinary and interdisciplinary"; Trench and Bucchi (2010, p. 4) defined science communication as a field that it is "not (yet) established as an academic discipline but that could emerge as a discipline with strong interdisciplinary characteristics or as a sub-discipline in the still-growing field of communication studies"; Gascoigne and Schiele (2020, p. 12) see it as "a young transdisciplinary field, still developing and evolving, but not yet regarded as a discipline".

It is striking that, over a twenty-year span, we see a repetition of the status of science communication as emerging. The thing that is science communication is always becoming. In Chapter 6, we look further at new dimensions and competing definitions of the field. The ways of thinking and speaking about science communication are in continuing flux.

In terms of its practice science communication in the first quarter of the 21st century has recognised but also contested roles. Political and societal awareness of science communication grew through the COVID-19 pandemic, when experts of many kinds played out their expertise in public. However, a common injunction of governments to their populations was to "follow the science". This prescribed a limited role for science communication as one of influencing populations to comply with government regulations and expert advice. But science communication played other roles even in this stressed context: explaining epidemiological modelling, narrating vaccine development and analysing public attitudes.

The experience of the pandemic strengthened for some the normative view of science communication as a means to build and boost public trust in science. This purpose was being increasingly attached to science communication in the years before the pandemic; this impetus came not only from institutions and organisations of the scientific communities but also from within science communication. We have dealt elsewhere (Chapter 3) with the underlying view of a public trust deficit. In the present context of science communication's consolidation as a thing, it is important to note that the norm ascribed to it as a trust-builder is also contested internally. In her keynote address at the PCST Network conference in 2023, Jahnavi Phalkey, Founding Dirctor of Science Gallery Bengaluru, asked the audience to consider if it was the science communicator's task to build trust in science. Science communicators, she offered, might perhaps consider themselves as independent professional practitioners who offer a perspective on scientific claims in context such that citizens are able to form informed opinions.

Several related processes can be observed in science communication's becoming A Thing:

- Institutionalisation – structures for science communication have been established within and alongside existing institutions of education, government, state, philanthropy and other non-governmental organisations;
- Professionalisation – education, training and career paths have developed, and continue to do so, for those taking up science communication as their principal employment;
- Internationalisation – formats, policies and programmes for science communication have spread and grown across all continents;
- Diversification – actors and platforms for science communication have multiplied, as we shall explore further in Chapter 5.

To these four processes we should add a fifth, although it is less pronounced so far than the others – commercialisation; opportunities have emerged and been created for private companies but also public bodies to sell expertise in science communication through contracts for training, advice and administrative and consultancy services. Some of these opportunities have arisen in the service circles around larger institutions, such as the European Commission, while others have been generated by start-ups taking a chance to find clients. This niche activity seems likely to grow in association with the continuing institutionalisation and professionalisation of science communication.

REFERENCES

Bauer, M. W., Gaskell, G., 1999. "Towards a Paradigm for Research on Social Representations". *Journal for the Theory of Social Behaviour* 29, 2, 163–186. https://doi.org/10.1111/1468-5914.00096

Bucchi, M., Fattorini, E., Saracino, B. 2022. "Public Perception of COVID-19 Vaccination in Italy: The Role of Trust and Experts' Communication". *International Journal of Public Health* 67, 1604222. https://doi.org/10.3389/ijph.2022.1604222

Bucchi, M., Trench, B., 2016. *Public Communication of Science*. London and New York: Routledge.

Durant, J., 1994. "What Is Scientific Literacy?". *European Review* 2 (1), 83–89.

Epstein, S., 1995. "The Construction of Lay Expertise: AIDS, Activism and the Forging of Credibility in the Reform of Clinical Trials". *Science, Technology and Human Values* 20 (4), 408–437.

Fleck, L., 1979/1935. *Genesis and Development of a Scientific Fact*. Chicago: University of Chicago Press.

Funtowicz, S., Ravetz, J. R., 1993. "Science for the Post-Normal Age". *Futures* 25, 735–755.

Gascoigne, T., Schiele, B., 2020. "Introduction: A Global Trend, an Emerging Field, a Multiplicity of Understandings: Science Communication in 39 Countries". In T. Gascoigne, B. Schiele, J. Leach, M. Riedlinger, B. V. Lewenstein, L. Massarani, P. Broks (Eds.), *Communicating Science: a Global Perspective*. Canberra: Australian National University Press, 1–14.

Gibbons, M., Limoges, C., Nowotny, H., Schwartzman, S., Scott, P., Trow, M., 1994. *The New Production of Knowledge: The Dynamics of Science and Research in Contemporary Societies*. London: Sage.

Goodell, R., 1977. *The Visible Scientists*. Boston: Little, Brown.

Habermas, J., 1971. "The Scientization of Politics and Public Opinion". In J. Habermas (Ed.), *Towards a Rational Society: Student Protest, Science, and Politics*. London: Heinemann.

Hornig Priest, S., 2010. "Coming of Age in the Academy? The Status of Our Emerging Field". *JCOM Journal of Science Communication* 9 (3), C06. https://doi.org/10.22323/2.09030306

Irwin, A., 1995. *Citizen Science; a Study of People, Expertise and Sustainable Development*. London: Routledge.

Kitty, A., 2018. *When Journalism Was a Thing*. London: Zero Books.

Kuhn, T., 1962. *The Structure of Scientific Revolutions*. Chicago: University of Chicago Press.

LeBard, R., 2020. What Jacinda Ardern can teach us about science communication – lessons from the Prime Minister of New Zealand during Covid-19. https://medium.com/scicomm/what-jacinda-arden-can-teach-us-about-science-communication-6f8fc42712b4

Logan, R., 2001. "Science Mass Communication: Its Conceptual History". *Science Communication* 23 (2), 135–163.

Merton, R. K., 1968. "The Matthew Effect in Science". *Science* 159 (3810), 56–63.

Miller, J. D., 1983. "Scientific Literacy: A Conceptual and Empirical Review". *Daedalus* 112 (2), 29–48.

Moscovici, S., 1961. *La psychanalyse, son image et son public*. Paris: Presses Universitaires de France.

Nelkin, D. 1995/1987. *Selling Science: How the Press Covers Science and Technology*, 2 editions, New York: W. H. Freeman

Nelkin, D., Lindee, S. M., 1995. *The DNA Mystique: The Gene as a Cultural Icon*. New York: W. H. Freeman.

Trench, B., 2023. "Strengthening Interdisciplinarity in Science Communication Education: Promise, Pleasures and Problems". *JCOM - Journal of Science Communication* 22 (6), Y02. https://doi.org/10.22323/2.22060402

Trench, B., Bucchi, M., 2010. "Science Communication, an Emerging Discipline". *JCOM – Journal of Science Communication* 9, 3.

Trench, B., Bucchi, M., 2021. "Global Spread of Science Communication: Institutions and Practices Across Continents". In M. Bucchi, B. Trench, (Eds.), *Routledge Handbook of Public Communication of Science and Technology*, 2nd ed. London and New York: Routledge, pp. 97–113

Wynne, B., 1992. "Misunderstood Misunderstanding: Social Identities and Public Uptake of Science". *Public Understanding of Science* 1 (3), 281–304.

5

ACTORS AND PLATFORMS
THE WHO AND WHERE OF SCIENCE COMMUNICATION

Interactions between art and science have become more common in recent times as part of the increasing variety of platforms of science communication. But such interactions can be found at the very beginning of modern science. Galileo Galilei had a background and training in the visual arts and this turned out to be essential to him recognising against the established dogma that the moon's surface was not smooth. His mastery of the technique of chiaroscuro contributed to his understanding of light and shadow on the moon as being due to cavities and irregularities. Galileo's friend, Cigoli, painted the moon in a depiction of the Immaculate Conception in 1612 as he understood it from Galileo's published observations in Sidereus Nuncius (Galileo, 1610).

Who communicates science and to whom? Originally, as we saw (Chapter 2) popularising scientists were addressing both colleagues and general audiences. Gradually, as science communication was thematised and prioritised also at the institutional level, professional communicators (journalists, popularisers, curators) became the main actors and gatekeepers of science communication. A relevant focus of science communication research and initiatives, particularly in the second part of the last century, was about addressing the limitations and shortcomings of the media (particularly journalism) in relaying science results and content to general audiences. Initiatives to address these limitations included training for science journalists and science media centres to assist them in better reporting about science. More recently, in response to the perceived shortcomings of the media in playing this role and to increasing pressure and incentives on research

DOI: 10.4324/9781032646749-5

organisations to communicate, the emphasis has swung back to scientists and research organisations addressing their non-expert publics directly.

Today, scientific institutions, visible scientists, universities and high-tech and pharma companies are increasingly active in science communication. This creates new opportunities, responsibilities, demands and expectations, for example in terms of recruiting internal competencies, training (both researchers and administrative staff) and evaluation. Institutions and organisations need both professional science communicators and a cohort of scientists active in public communication. Strategies for supporting and rewarding such communication need to distinguish between full-time and part-time, vocational and voluntary doers of science communication.

ACTORS OF CONTEMPORARY SCIENCE COMMUNICATION

Professional Communicators

For those in this category, professional education, continuing professional development and recognisable career paths are required. Those paths are not clearly delineated, and the content of the job, science communicator, can vary widely – and is subject to multiple influences, including recent technological developments (e.g. AI). Their mission is normally focused more on institutional communication and related strategy and visibility, and to this aim, they can work with individual scientists to support them in the communication of their research and expertise. The communities of professional science communicators are notably networked and internationalised – including in job mobility – and they are predominantly female, in ways that are often commented on but not well understood.

In Chapter 4, we described how the establishment of professional education was a mark of science communication "becoming a thing". The spread of postgraduate programmes and continuing professional development courses has been a mark beyond that of the professionalisation of science communication as an employment niche. Almost certainly, there will continue to be a path for those with science degrees, maybe early-career research experience, but also a personal aptitude for public communication, to find positions as professional science communicators. It would be unreasonable to expect, or insist, that all those entering this field have a formal qualification in science communication but such qualifications are offered across increasing numbers of educational and cultural contexts and are often desiderata in job specifications.

Scientists as Communicators

Scientists can be involved in science communication for different reasons, in multiple diverse ways, and with variable degrees of intensity. Reasons to become active in science communication may include the genuine desire to share their results and knowledge with the broader community, to contribute to the visibility and recognition of their own institution affiliation and to respond to institutional or funding agencies' demand for "dissemination". At the more personal level, they may enjoy personal visibility or in some cases even celebrity. Modes of communication can vary widely: interviews in general news media, public talks, engagement with different social media.

The intensity of engagement covers a broad spectrum from contributing to communication activities planned at the organisational level (open days, festivals, science cafes) to more active, in some cases quasi-entrepreneurial communication activities (popular talks and performances, authorship of articles; videos on social media; regular contributions to TV programmes). Public communication activities are normally not remunerated by research organisations to which scientists are affiliated, nor they are recognised as part of the standard activities. Some activities can, however, be remunerated by media outlets or by event organisers and sponsors. These intensified efforts of individual scholars do not always pursue the same goals as the scientific organisations they are working for or affiliated with. Individual and organisational goals in communication can be misaligned, from being slightly at odds to being diametrically opposed.

These tensions can be further catalysed by other factors such as journalists' sourcing practices: journalists often bypass the organisational press and communication departments to call researchers directly, often to pick faces and voices of scientists that are already familiar to the public and are familiar with media demands, but sometimes also to get less generic statements and even feed controversy as a news value. News media and institutional communication teams may have diverging views on who constitutes a suitable public spokesperson: While institutions might favour topical competence and institutional loyalty, journalists might look more for charisma, appearance, readiness to respond, etc. (Peters, 2021). The "Matthew Effect" discussed in Chapter 4 may exacerbate this even further: as the visibility of certain scientists builds up, it can incentivise both the scientists themselves and the journalists to expand the areas and topics on which the scientists are invited and willing to comment – not because they necessarily have a pronounced competence in the respective field, but because they have become familiar figures and faces who are able to express themselves in ways that are media-genic. The personal motivations of individual scientists may be different again – they may want to raise their own public profile and gain advantages due to prominence, increased visibility of their work among organisational leaders and peers, and rises in their citation counts (Fahy and Lewenstein, 2021).

Unlike professional communicators, voluntary science communicators are dispersed, defined much more by their institutional affiliation than by their occasional communication activity. A long-term historical analysis of scientists' shifting involvements in science communication would be very valuable. Economist Albert Hirschman (1982) introduced the concept of shifting involvements to describe what he saw as an oscillation of advanced societies between focus on private well-being and engagement in public life and discussion. A similar approach could be adopted to analyse the oscillation of scientists from intense personal involvement in popular science communication (particularly from the mid-19th century to the mid-20th century) to a more moderate engagement that corresponded to a substantive growth of the role of mediators; and then back again to the front-stage of science communication.

PLATFORMS FOR SCIENCE COMMUNICATION

Besides public talks, lectures and demonstrations, traditional platforms for science communication included popular books, magazines and newspapers, as discussed in Chapter 2. Although some institutions (e.g. the Royal Institution in London) recognised public communication as part of their mission, most of the activities were partially, if not entirely, supported by paying audiences. A commercial model continued to operate in the 20th century, with some changes in the relevance of different media, in particular, the rise of television. At the same time, not only research institutions but also public broadcasters started to recognise the popularisation of science as part of their duties.

Changes in the media landscape and business have again changed the model for supporting science communication. Traditional media industries increasingly struggle to get readers to pay for content –

including science-related content. On the other hand, key contemporary communication business players like Google (owner also of YouTube) and Meta (Facebook and Instagram) are not content producers or editors. With few exceptions, the revenues available through these channels are not enough to sustain a professional science communicator. So, eventually, most contemporary public science communication is driven and paid for by public institutions, either directly or indirectly (for example, funding research projects with requirements for their dissemination, as in the case of European Commission funding schemes and several funding schemes worldwide). This shift has consequences at several levels, many of which still need to be understood, including the issue of the quality of science communication. In the following series of short sections, we present notes on some of the many, and proliferating, platforms for science communication, both deliberate and accidental. Several of these have long and continuing histories but here we indicate their present positions and possibilities.

Books

Popular books are the most traditional platform for communicating science to the public, as we saw earlier (Chapter 2). As Bruce Lewenstein noted (2009), a shift occurred in the 1970s, with popular science books becoming a regular feature of bestseller lists and book prizes. A landmark moment is usually recognised in the publication of *A Brief History of the Universe* (1988) by leading astrophysicist Stephen Hawking, a demanding read which became a top-seller worldwide and contributed to establishing its author as scientific celebrity. Relevant examples in this century include *Seven Brief Lessons on Physics* (2014) by Carlo Rovelli and *Thinking, Fast and Slow* (2011) by Nobel laureate Daniel Kahneman, an uncommon example of the successful popularisation of the cognitive sciences.

Our view of science communication as conversation suggests not limiting attention to popularisation books strictly defined, i.e. essays aimed at popularising a scientific branch, topic, or figure. Books relevant to the conversation about science could be broadly classified as follows: Essays dedicated to a scientific sector, topic or personality; essays on other topics (e.g. management, history, etc.) that touch upon scientific content; fiction centred on science topics or figures; fiction that touches on scientific content (see the following section).

All of these categories can include graphic/comic books about science and scientists, a genre which has become increasingly prominent in the past few decades, e.g. Logicomix (Doxiadis, 2009), about philosophy of science; a graphic biography on Richard Feynman (Ottaviani and Myrick, 2011). In Japan, comics (manga) played a key role in the heated debate after the Fukushima accident of 2011; particularly notable is the graphic memoir by a technician at the power plant, Kazuto Tatsuta (2017).

Science (in) Fiction

We mentioned science fiction's place in the history of science and its connection with science popularisation earlier (Chapter 2). The depictions of science and of scientists in this genre have been much studied but its role in influencing scientific effort and the study and career paths of scientists deserves more analysis. Away from this genre's frequently dystopian speculations about the future, some novelists have looked to historical figures in science as literary characters. Serial prize-winning author John Banville launched his literary career in the late 1970s and early 1980s with historical novels on Copernicus, Kepler and Newton; his more recent work includes crime novels with a forensic pathologist as a central character. Other crime writers have also dug into forensic science, both historical and contemporary, as part of their story-telling.

Chemist and novelist Primo Levi made characters of the chemical elements in his autobiographical work, *The Periodic Table* (1975). Michael Crichton's medical education informed his novel writing; some of his 26 novels were cautionary tales of scientific or technological hubris, perhaps most famously *Jurassic Park* (1990), about genetic engineering, but also *Prey* (2002), about nanotechnology. His cautions extended to climate science in *State of Fear* (2004), which made Crichton a very controversial figure. Other commercially successful novels have featured scientists as central characters, such as *Intuition* (2006) by Allegra Goodman, which is set in a cancer research lab, and *Lessons in Chemistry* (2022) by Bonnie Garmus, which is largely set in a chemical research institute.

Newspapers

Newspapers are traditional platforms for science communication, featuring news, comments, columns, stories and interviews about science and scientists. Many studies have documented long-term trends and changes in the coverage of science by the daily press in, for example, Britain, Bulgaria, Italy, Greece, Canada and Spain. In terms of topics, the physical sciences long dominated this coverage, at least until the last decades of the 20th century, when the life sciences and medical sciences became more relevant.

In that period also, some newspapers started to dedicate to science a specifically framed section, normally published on a weekly basis. This sparked a debate in science communication studies about the so-called "ghettoization" of science news by having it in a dedicated section rather than distributed across the newspaper. Higher accuracy of reporting seems one advantage of dedicated sections, where science is generally covered by science journalists, rather than by generalist journalists, balanced by the lower impact due to the self-selection of readers who actually read such sections. Dedicated science sections, sometimes with links in their titles to medicine, environment or climate, continue to be published in "up-market" newspapers around the world. *Le Monde*, in France, for example, publishes a Thursday "cahier" (notebook) with analytical features, commentaries, columns, book reviews and profiles on issues, trends and personalities in science and medicine.

In many ways, it is when science becomes general news, featured alongside politics, current affairs and economy, that implications become more relevant and interesting for culture and society (as well as for science itself) and more likely to contribute to social conversations. In the COVID-19 pandemic, this happened on a grand scale, with explanatory articles and infographics on viruses and vaccines, though also frequently triggering the debate about the supposed mismatch between generalist reporters and complex scientific information.

Magazines

A similar dualism to that in newspapers can be found also in science coverage in magazines. Science can be covered in dedicated or general sections by generalist magazines or magazines explicitly devoted to popular sciences. Among the latter category, *Scientific American* is perhaps best-known; interestingly, it and other periodicals spread internationally, also through national editions which maintained their own autonomy and even different titles (*Le Scienze* in Italy; *Spektrum der Wissenschaft* in Germany; *Global Science* in China). The same happened with the German magazine *Focus*, published in several European and Latin American countries with different content and topics, and with the massively popular *National Geographic*. The British magazine *New Scientist* has a wide international reputation.

Printed magazines about science experienced a similar decline in circulation figures to that of newspapers and general magazines, particularly in the 21st century, in connection with changes in media consumption and information consumption habits. Some magazines invested heavily in their online presence, including creating opportunities for long-form articles, and for blogs by high-profile writers. At the same time, new publications emerged in online form only, such as *Nautilus*, "a different kind of science magazine … [whose] stories take you into the depths of science and spotlight its ripples in our lives and cultures".

Public Talks

The long tradition of public talks popularising science continued into the 20th century when acclaimed international popular lecture tours in Europe, Northern America and Japan contributed to establishing Albert Einstein as one of the greatest celebrities of the time. The rise of television made public talks less relevant but in the 1990s a new interest was spurred by public institutions as part of their efforts in communicating science. In 1993, for example, the European Commission launched the European Science Week, soon followed by European Researchers' Night. Backed also by national and international funding, new festivals and events were created or existing initiatives revitalised. Science Festivals like those in Edinburgh, New York or Genoa became a regular attraction for hundreds of thousands of participants, offering opportunities to both leading and emerging scientists. The business model for these events has been generally a mixed one, with a combination of financial support from private and public sponsors and paying audiences. In some cases, research organisations pay a fee or cover expenses to showcase their research. Scientists' participation in such events is normally on a voluntary basis, but in some cases, they may also receive a fee.

The business aspect of public science talks becomes prominent when it comes to lecture tours by highly visible scientists, typically known more for their television presenting or writing of books than for their scientific work, if, indeed, they still do any. Brian Cox is a particle physicist with a busy career in radio and TV, where he appears in various roles, and he commands large audiences at high ticket prices with his tours to theatres and other venues usually hosting rock concerts. Richard Dawkins undertakes such tours around the publication of his latest book, again selling out large venues with his frequently provocative presentations.

Audience analysis and evaluation of the impact of these initiatives has not been systematic but we do know a few things about such events: participants are largely self-selected in terms of educational level and interest in science, compared to the general population; surveys point to direct encounter with scientists as a highly rated occasion in terms of perceived credibility by audiences (compared for example to TV or social media content); impacts on audiences' knowledge are not well documented, but relevant impacts can be found on participating scientists, e.g. in terms of challenging stereotypes about popular audiences or fostering motivation to engage in future communication activities. These communication experiences often also reveal to scientists and their organisations the need for better preparation and training.

Short Talks and Competitions

Early-career researchers, in particular, are increasingly called on to present their work in compressed packages to general audiences, sometimes as part of festivals or open days, but also in everyday settings that they control themselves and in competitions run by institutions or agencies. One of the best-known of these latter formats was Fame Lab, in which researchers aimed to engage audiences with their projects through short, vivid presentations that were judged competitively for their impact. The format originated in Britain but was applied in national contests in many countries, with an international champion selected through a competition of national winners. Particularly popular in Germany and Russia, but also being taken up in Turkey and the United States, are Science Slams, organised along similar lines with regional competitions of 10-minute presentations producing contestants for national competitions. Ma Thèse en 180 Secondes is a French version of the short pitch, judged within and between institutions, and also featuring national and international competitions across increasing numbers of francophone countries. Threesis is an English-language version of the same idea. Pint of Science, as the title suggests, is run in bars and pubs and originated in Britain in 2013 but has spread to many more countries, with a first global festival in May 2025; speakers are constrained not so much by time but by making themselves understood and heard over other distractions. Some of these initiatives have or have had training associated with them and the competitive element can be the source of valuable feedback. For some individuals, participation in events like these has led to changes in career direction.

Science Cafés

Based on the French tradition of *cafés philosophiques*, the *café scientifique*, as it was known long-time even outside francophone contexts, involves presentation and – importantly – discussion of scientific information and ideas in cafés and similar everyday settings. From the 1990s the format has spread to many countries and has been adapted to local circumstances. The common element across multiple versions of the format is the invited expert or experts presenting an idea, their current research, or their lives in science. As befits the name and setting, the intention is to encourage conversation between experts and non-experts, which the invited guests often find challenging but stimulating. In the context of science communication as conversation, the science café is a privileged example, where the host organisation and the presenter(s) are – in the best circumstances – open to the discussion going in unanticipated directions.

Exhibitions and Museums

Exhibitions featuring science and technology in the late 19th and early 20th centuries had a strong focus on innovations and advances. These large-scale displays of objects have given way to more mobile exhibits that are often part of events like science festivals, as mentioned above. Over recent decades, as part of the drive for engagement and public communication, research institutions worldwide have hosted exhibitions on their own premises, often explicitly commissioned to popularise their research. Anniversaries of discoveries or of the births or deaths of renowned scientists are also occasions for thematic exhibitions.

Science museums and science centres are the established venues for such exhibitions. The typical arrangement features a combination of the museum's own collections or installations and temporary exhibitions, the latter being planned with a view both to attract occasional visitors that would otherwise not visit a science museum and to encourage repeated visits. Most science centres and museums have to balance external support and revenues from visitor tickets and restaurants/shops. An interestingly different model was launched in 2008 with the Science Gallery Dublin, later exported and adapted to other places, including Melbourne, Bengaluru and Detroit. Unlike most science centres, the Gallery did not have any permanent collection or exhibition; instead, it offered visitors open access to thematic, interactive exhibitions based on creative collaborations between scientists and artists (see Figure 5.1).

Figure 5.1   An exhibit on the Coarse Fabric of Being Human, part of Carbon at Science Gallery Bengaluru, 2024. Photo courtesy of Science Gallery Bengaluru, India.

Television and Streaming Platforms

Early in the history of television as a mass medium, science popularisation programmes were launched in several countries. The main format was originally the documentary about science, technology, or natural history topics. In the most successful cases, the unifying and recognisable narrative continuity was provided by a presenter with expert knowledge. Scientists and popularisers like David Attenborough or Carl Sagan became familiar figures to television viewers worldwide. Some documentaries were produced by leading production companies, like Disney, as animated cartoons, aimed at children and family audiences. During the 1950s, Disney produced a series of animated cartoons about science and technology topics that were aired by ABC network. Topics included atomic energy (*Our Friend the Atom*), space exploration (*Man in Space, Man and the Moon, Mars and Beyond*), mathematics (*Donald in mathmagic land*). Visible scientists like Werner von Braun acted as consultants for some of these productions and also appeared in the films. The cartoons had great success in terms of audience and a remarkable influence in building awareness of the importance of space enterprises and investments within American public opinion.

In Britain, the BBC has broadcast the *Horizon* programme since 1964. In Italy, *Quark* (later *Superquark*) has been aired in prime-time RAI1 since 1981, with audience peaks of more than 5 million, establishing its presenter Piero Angela, a journalist with no formal science training or degree, as a celebrity and spokesperson for science. In the United States, public television PBS has broadcast *NOVA* in prime time since 1974.

Some of the most popular TV fiction series of the last decades, like *Big Bang Theory* (2007-2019) or *The Simpsons* (1989-present) feature relevant references to scientific content and scientific figures. With the advent of on-demand TV platforms like Disney+ or Netflix, productions – both fiction and documentaries – about science and related topics have been newly produced or relaunched.

Radio and Podcasts

Before the diffusion of television, radio played a major role in several areas of information and entertainment, including science and related topics. Up to now, radio programmes about science continue to be broadcast regularly by both public and private radios. Scientists have often contributed to these programmes: compared to television, radio does not involve big production costs, audience numbers are less important and scientific guests can easily be interviewed by phone. Nowadays, audio programmes about science are often listened to in the form of podcasts, i.e. on demand, through smartphones and other devices. Content available includes digital versions of broadcast programmes as well as dedicated productions, sometimes hosted by individual scientists or well-established science communicators. The business model may involve sponsorship by companies, foundations, public bodies, together with commercial advertisements. In many of these contemporary productions, the boundaries of "science communication" are interpreted rather broadly, including technology advances, related cultural topics and political issues.

The growing population of podcasts focusing on science includes those hosted by visible scientists and those hosted by journalists, for example, as adjuncts to radio programmes or other traditional media outlets. The relatively informal character of podcasts and their freedom from the constraints of a radio schedule means they can facilitate extended conversations of a kind not generally found in any other media platforms.

Greater breadth can be found in magazine-type radio programmes where science and science-related content may feature on the basis, for example, of a recently published popular science book or a piece of research. Interviews by presenters without a science brief can go to places that may discomfort scientist interviewees but also reach larger audiences than those for science-nominated radio productions. Those larger numbers are also accessible to scientists and science communicators who have a regular "slot" on current affairs or magazine programmes, where again they need to modulate their contributions to the broader publics.

Social Media

Since the advent of Facebook in 2004, social media have been used by scientists, scientific organisations and popularisers to communicate science. The term "social media" obviously underscores a variety of content and approaches to science communication. Several science organisations, for example, have Facebook, X (formerly Twitter) and Instagram pages to update their potential audiences about events and new achievements. Individual scientists use X to signal new publications and events; during the pandemic, some of them also extensively used it to publish their comments and analyses of epidemiological data. LinkedIn is also widely used for networking and updates.

YouTube features highly popular channels by "science youtubers": typically, a young person with a scientific background who addresses broad science-related topics in an informal, dynamic manner during short videos. Some of these, like former NASA engineer Mark Rober, have tens of millions of followers and views. Unlike other social platforms, YouTube pays the most popular content producers revenue, according to certain criteria. However, studies have highlighted that for most of them, it is not easy to turn this activity, often highly time-consuming, into a viable, remunerative professional career. Likewise, the more recently established video channel TikTok features several science popularisers with a few million followers. For many of these "science influencers", making content for these channels feeds into other activities like public talks, popular books and TV appearances, contributing to the increasing "hybridisation" of science communication content across different media and platforms.

Cinema

Science and scientists in cinema have been studied by science communication scholars in terms of content analysis, looking for example at dominant topics in different historical periods that to some extent reflect – and contribute to feed – societal attention, debate and concern with regard to certain areas of science (genetics, nuclear science, space science) and related issues (gender and science). Another established area of study has been on representations and stereotypes of scientists as portrayed in movie productions, e.g. the helpless scientist, the mad scientist, the amoral scientist, the absent-minded professor, the heroic scientist (Kirby and Ockert, 2021). Christopher Nolan's 2023 film about Robert Oppenheimer, the scientific coordinator of the project leading to the first atomic bomb, based on a book biography, had great and unanticipated commercial success and several Academy awards.

Examples of scientists' work as advisers to film productions feeding back into original research, as mentioned in Chapter 2, continued into our times. Astrophysicist Kip Thorne advised director Christopher Nolan for his acclaimed *Interstellar* (2014). Work needed to accurately represent black holes in the movie according to state-of-the-art science led to an original paper by Thorne and colleagues, "Gravitational Lensing by Spinning Black Holes in Astrophysics, and in the Movie Interstellar" (James, von Tunzelmann, Franklin and Thorne, 2015). Thorne received the Nobel prize for physics in 2017 for his work on gravitational waves. Mathematical simulations by scientists at the University of California for representing different states of snow and ice for the Disney film, *Frozen* (2013), also led to original publications in mathematical journals.

Comedy

While humour has been long used as part of telling stories of science in journalism and public talks, comedy around science is a more recent and challenging proposition. If a joke is based on an assumed understanding or familiarity with a scientific idea, it may well fall flat with a broader audience. However, some stand-up comedians have drawn on their scientific interests and/or qualifications to develop whole shows or parts of shows around topics relating to science. Also, some public communicating scientists have specialised in comedy as part of their routine; German physicist Sabine Hossenfelder was a researcher in several countries for twenty years and is a popular science writer, blogger, and contributor to magazines, as well as having a YouTube channel with over one million subscribers where she explores scientific ideas through a comic lens. Climate scientist John Cook has developed a comic character, Cranky Uncle, as part of his climate change communication. British-based television presenters Dara O Briain – a comedian qualified in physics – and Brian Cox – a particle physicist and science celebrity – approach the explanation of natural, particularly cosmological, phenomena through informal, joking banter between them.

Others working in science and comedy have offered their services to companies and institutions as trainers in telling science stories with humour. Hello SciComm seeks to "combine science communication and comedy", providing training, advice and jokes, including for National Geographic, Stanford professors and Barack Obama.

Comic elements are often included in the kinds of short talks referred to earlier, and in the case of the Bright Club, comedy is central. Here, researchers are challenged to talk about their work in a comedy club setting. The initiative started at University College London in 2009, and the format has been taken up in many cities and institutions across Britain, Ireland and Australia.

Theatre

Galileo Galilei, whom we have met in other contexts, comes up here again, as the central character in Bertold Brecht's play *Life of Galileo*

(1938), just one of several notable 20th-century plays in which the sometimes-dramatic lives of scientists (in Galileo's case, very dramatic) are represented on stage. Earlier in the century, Bernard Shaw explored the conflicts between private medicine as a business and the moral obligations of the doctor in *The Doctor's Dilemma* (1906). Moral and other dilemmas are also integral to Michael Frayn's *Copenhagen* (2000), which reconstructs imaginatively the tense wartime conversations between Danish physicist Niels Bohr and German physicist Werner Heisenberg. These examples derive their drama from the competing demands on scientists in political and commercial contexts. Controversies within science are also material for dramatic representation.

As with other cultural forms, theatre has been exploited as a medium for public engagement with science. Science festivals have long included elements of theatrical performance. CERN, the particle physics research centre, hosted science and theatre events for some years and the work at CERN inspired a solo show by scientist-engineer Niamh Shaw, *That's about the size of it* (2011), which prompted an invitation to her from the institution to cooperate with their Arts@CERN programme. Shaw's five later shows all involved dramatic engagement with aspects of science.

The genre of science theatre has grown to the extent that it now has its own festivals, including that at Heilbronn, Germany, in association with the local science centre, Experimenta, since 2021, and another in New York, that is women-driven, running over three days with three plays and short talks, also since 2021. Scientists, actors and directors active in science theatre contributed to a collection of essays on the topic (Weitkamp and Almeida, 2022) which represented a coming-of-age of the genre and which made the harnessing of theatre explicit in its sub-title, Communicating Science and Technology through the Performing Arts. Issues examined in the chapters included definitions, institutions involved, participants, topics, venues and formats.

Poetry

The writing of poems inspired by scientific ideas go back at least as far as the early days of modern science. Scientists as well as poets – indeed, some scientist-poets such as the Czech immunologist and poet Miroslav Holub – have contributed to the genre, neatly encapsulated in a collection of 101 poems about science (Riordan and Turney,

2000). The collection's title, *A Quark for Mister Mark*, is adapted from a phrase in James Joyce's *Finnegans Wake* that gave the name adopted for a specific sub-atomic particle. The connections between science and poetry have also been explored – as with fiction and popular science writing – in reference to the use of analogy and metaphor in both, and in reference to the aesthetic appeal of science expressed as "the poetry of science". Richard Dawkins has adopted this phrase, having also written *Unweaving the Rainbow* (1998), with a title taken from the English 19th century poet John Keats, who was concerned that science was taking the magic out of nature; Dawkins's aim was to reassure Keats and those like him.

Geoscientist and science communicator Sam Illingworth has written on poetry's influence on selected scientists (2019), including Humphry Davy, a friend of the poets Wordsworth, Coleridge and Byron (see also Chapter 2). As with other cultural forms, poetry has been harnessed to science communication and Illingworth's work-shops help participants to write a poem around science, or at least to start the process. Illingworth also edits a journal, *Consilience*, that publishes poetry and artwork about science, borrowing a title from entomologist E. O. Wilson (1998) on the "unity of knowledge" and the possible "jumping together" of sciences and humanities.

Non-verbal Science Communication

A scientist preparing their presentation for a conference is likely to include graphs, charts and images that will tell their research story in a concise and compelling way. Drawings and diagrams have been a key part of scientific communication among peers for centuries, even to the point that fabrication or manipulation of images has become a recognised aspect of scientific fraud. In their use of images, astronomer Galileo, anatomist Vesalius and microscopist Hooke – mentioned else-where in this book – had a wider lay audience in mind.

Introducing a comprehensive collection of historical studies on non-verbal science communication, Mazzolini (1993) lists six main groups of sources for such history:

1. scientific instruments in all their variety and uses;
2. models, such as models of the planetary system, of the earth, of the human body (made in materials like clay, wood, ivory, wax, coloured *papier-maché* and plastic), of geological strata, of mo-lecular structures, of machines and so on;

3. scientific illustrations of plants, animals, anatomical dissections, geographical regions, of the sky, of planets;
4. classified collections of natural objects such as plants (herbals), minerals, living and stuffed or otherwise preserved animals, collections of anatomical specimens (whether healthy, pathological or monstrous), collections of bones and skulls;
5. arenas of scientific communication and practice such as natural history museums, botanical gardens, laboratories, anatomical theatres, and observatories, where a functional architecture provides the physical conditions for specific forms of communication;
6. conventional representations including graphs and symbolic artefacts such as classifying schemes, graphic representations of laws, diagrams, tables, and other varieties of non-verbal symbolism which are present in a wide range of disciplines.

We could further add to this list photographs and images obtained through different types of machinery (from X-rays to contemporary astronomical images and representations), also monuments and sculptures dedicated to scientists, which have played an important role in shaping the public image of scientist, particularly since the 19th century. Obviously, some of these sources work in combination with verbal descriptions and comments.

In the modern era, with the development of photography, television and digital media, the visual part of science communication has become more varied and more important. Photographer and scholar Felice Frankel has been working for decades with scientists to help them visualise their results and experiments (Frankel, 2005; see Figure 5.2). Her images have been featured on the cover of leading journals like *Nature* and *Science*. "I want researchers to understand — especially the next generation coming up — that the process of communicating the essential pieces of your research is not just about making pretty pictures, but about advancing your own thinking, clarifying your own work in a way that's accessible to non-experts".[1]

The contemporary professional science communicator needs to be conversant with the possibilities of Instagram and TikTok, as well as with those of animated images and infographics. During the COVID-19 pandemic infographics representing viral transmission and vaccine development were part of the repertoire for setting out basic facts. Microbiologist Siouxsie Wiles has worked with artists on films and installations on microscopic organisms, and in the early stages of the pandemic, she collaborated with cartoonist Toby Morris to

Figure 5.2   Yeast Flower, © Felice Frankel. Image reproduced by kind permission of the author.

produce an animated GIF comic, Flatten the Curve, that was widely circulated around the world[2].

Art-Science Collaborations

Many historical examples exist of works by renowned artists that draw upon and (re)elaborate scientific results and ideas. A relatively new phenomenon in recent decades is the promotion of art–science collaboration as a means intended to provide new and inspirational ways of communicating science. In this context, for example, science institutions have encouraged artists' residencies as part of their activities, for example, when visual and sound artist Laurie Anderson became officially the first "artist in residence" at NASA. Nowadays, science organisations worldwide host, encourage and sometimes sponsor artworks, installations and performances that connect in different ways with their research. The aims and expectations of these activities are not always clear and explicit: they may be intended as generally enhancing the profile and visibility of the organisations, attracting media attention and new types of visitors but they may also arise from artists' authentic interest in exploring in creative formats the implications of scientific work, such as a sculptor commissioned to provide a large piece of public art who drew on his study of "artistic responses to cosmological phenomena" such as the presence of dark matter to depict The Presence of an Absence[3].

Museums and exhibition spaces have also promoted and hosted projects and works inspired by science and based on collaborations between artists and scientists. Reviewing this spreading activity Halpern and Rogers (2021) propose a typology of art-science collaborations:

a.   Conveyance, which conveys a direct message, for example popularising a certain science topic;
b.   Contributive, which generates new knowledge;
c.   Contextual, which emphasise the context of science production and application
d.   Critical, which raises critical issues and questions about science developments.

Some of those involved in such collaborations refer to them as examples of STEAM, adding an A for Art to the established STEM acronym (see Chapter 4). In doing so, they harness art to the highly instrumentalised conception of science communication as serving the development of science and technology as economic drivers. Halpern and Rogers (ibid.) caution eloquently on this trend; the autonomy of artistic endeavour needs to be protected. Lévy-Leblond writes in the same spirit, presenting his 2011 treatise with a double title that requires only a few letters to be changed to change its meaning completely: *La Science et L'Art* / *La Science n'est pas l'Art* (Science and Art. Science is not Art).

HYBRIDISATION AND DIVERSIFICATION OF ACTORS AND PLATFORMS

The platforms and formats of science communication outlined above obviously mix, feed into and hybridise each other, following an increasing trend of contemporary communication, not just with regard to science. For example, science communicators and scientists who become familiar to audiences through television or social media can be invited to give public talks, write books or columns for newspapers or magazines.

Traditional formats can also be revamped, revitalised, adapted to new platforms. Public talks, for examples, found a new popularity through TED talks or similar formats being made available in social media and streaming channels. Also, recent popular formats of short presentations by young scientists that may have initially been intended as "elevator pitches" or presentations in bars or cafés of their work suit well further circulation through social media.

In a vision of science communication as conversation, the range of actors further diversifies, beyond professional and occasional to accidental science communicators, including for example artists, writers, moviemakers – eventually everyone is potentially an actor in communicating science.

The content and forms of science communication have diversified alongside the broadening of the range of actors. Typical units in the history of science communication were the public or popular lecture (e.g. at the Royal Institution) and the "story" as media news format. Stories about science in the media can be exposition of facts, interpretation of findings, celebration of discoveries or narration of processes. A proposed classification of science stories (Trench, 2020) includes these types, with their associated representation of science:

- Explanation and Profile – science is represented to society as a repository and continuing source of established fact
- Discovery – science as supplier of recently-found knowledge or solutions
- Promise – science as source of near-future applications and solutions
- Celebration – science as a place of outstanding achievements
- Caution and reassurance – science warning about developments in society or nature, or conversely reassuring about such perceptions
- Controversy and Competition – science as a place of disputes and conflicts
- Interrogation – science producing information and ideas to be critically examined
- Speculation – science offering insights of far-future developments

At the core of media coverage of science are science stories. Journalists need to engage their readers, viewers and listeners with the material they select as newsworthy and they do this by means of story-telling and argument. In media reporting and analysis, there is almost always an event or events, a character or characters, arranged in a narrative. "Journalists are [the] professional story-tellers of our age ... the journalist's work is focused on the getting and writing of stories" (Bell, 1991, p. 147).

Looking at cultural exchanges more widely, we can see that stories are everywhere, they may be considered the bedrock of culture. In advocating a "cultural approach to science communication", Davies et al. (2019) write that such an approach "acknowledges that public stories about science (as we understand science communication) should not be construed as fundamentally different to other kinds of public story-telling, but as both intertwined with, and bearing marked similarities to, public culture and entertainment". This paper's title opens with "Stories as Culture"; the approach outlined involves understanding and interpreting such stories as representing the experience people have of science, and the meanings they assign to that experience.

In this perspective, focusing on stories is a valuable means of making sense of science in society and culture. But, as with so many other cultural phenomena – including many of those described above – a larger part of science communication effort involves turning them into instruments. Thus, storytelling becomes a key part of training and strategising in science communication. In preparation for public talks, communicating scientists are encouraged to figure out how their research can be told as a story, with a narrative path around a problem and a solution, maybe with objects or phenomena imagined as characters. The central character is often the scientist herself solving a problem or confronting adversaries and overcoming dangers (Dahlstrom, 2014).

Similar discussions of the structures and purposes of science storytelling were published alongside the article by Davies et al these included in *JCOM – Journal of Science Communication*, 2019, no. 5; these included a treatment by Joubert, Davis and Metcalfe (2019) of storytelling as "the soul of science communication". León (2024) introduces another set of such articles by saying that "we know that the story can act as the glue that creates a unified and continuous statement, making the audience more likely to engage with and remember the knowledge being communicated" (p. 132). But he also draws attention to the risks of distortion and oversimplification in story-telling and, indeed, in all science communication. Three Mexican science journalists (Rueda, Rosen and Crúz-Mena, 2024) work their way through published studies of science story-telling in search of definitions, emerging to focus on the challenges of associating story and meaning. The storyteller's intended meaning may differ from that given by the recipients; the authors, for their part, say that meaning "is given by the level of understanding of the arguments presented" (p. 157). Thus, Rueda et al give a quite different emphasis to meaning-making from that espoused by Davies et al, adding to the plurality of views – even around central ideas – that marks the field of science communication.

NOTES

1. https://news.mit.edu/2023/eyes-have-it-frankel-1204.
2. https://thespinoff.co.nz/media/07-09-2021/the-great-toby-morris-siouxsie-wiles-covid-19-omnibus
3. Sculptor Lar O'Toole describing his ten-metre-high steel structure in a Dublin office complex (The Wexford People, 29 May 2024).

REFERENCES

Bell, A., 1991. *The Language of News media*. Oxford: Blackwell.

Dahlstrom, M. F., 2014. "Using Narratives and Storytelling to Communicate Science With Nonexpert Audiences". *PNAS Proceedings of the National Academy of Sciences of the United States of America* 111 (4), 13614–13620. https://doi.org/10.1073/pnas.1320645111

Davies, S. R., Halpern, M., Horst, M., Kirby, D. A., Lewenstein, B., 2019. "Science Stories as Culture: Experience, Identity, Narrative and Emotion in Public Communication of Science". *JCOM* 18 (5), A01. https://doi.org/10.22323/2.18050201

Dawkins, R., 1998. *Unweaving the Rainbow: Science, Delusion and the Appetite for Wonder*. Boston: Houghton Mifflin.

Doxiadis, A., 2009. *Logicomix: An Epic Search for Truth*. London: Bloomsbury.

Fahy, D., Lewenstein, B., 2021. Scientists in Popular Culture: The Making of Celebrities. In M. Bucchi, B. Trench (Eds.), *Routledge Handbook of Public Communication of Science and Technology*. London: Routledge, 33–52.

Frankel, F., 2005. *Visions of Science*. Milan: Olivares.

Galileo, G., 2008/1610. Sidereus Nuncius. In *Opere*, vol. I, Milan: Mondadori.

Garmus, B., 2022. *Lessons in Chemistry*. New York: Knopf Doubleday.

Halpern, M. K., Rogers, H. S., 2021. "Art-science Collaborations, Complexities and Challenges". In M. Bucchi, B. Trench (Eds.), *Routledge Handbook of Public Communication of Science and Technology*. London: Routledge, 214–237.

Hirschman, A. O., 1982. *Shifting Involvements: Private Interest and Public Action*. Princeton, NJ: Princeton University Press.

Illingworth, S., 2019. *A Sonnet to Science. Scientists and Their Poetry*. Manchester: Manchester University Press.

James, O., von Tunzelmann, E., Franklin, P., Thorne, K. S., 2015. "Gravitational Lensing by Spinning Black Holes in Astrophysics, and in the Movie Interstellar". *Classical and Quantum Gravity*, 32 (6). https://iopscience.iop.org/article/10.1088/0264-9381/32/6/065001

Joubert, M., Davis, L., Metcalfe, J., 2019. "Storytelling: The Soul of Science Communication". *JCOM* 18(05), E. https://doi.org/10.22323/2.18050501

Kirby, D. A., Ockert, I., 2021. "Science and Technology in Film: Themes and Representations". In M. Bucchi, B. Trench (Eds.), *Routledge Handbook of Public Communication of Science and Technology*. London and New York: Routledge, 77–96.

León, B., 2024. "Introduction: Science through Stories". *Mètode Science Studies Journal* 14, 132–135 https://metode.org/issues/monographs/introduction-science-through-stories.html

Lewenstein, B. V., 2009. "Science Books Since World War II". In D. P. Nord, M. Schudson, J. Rubin (Eds.), *The Enduring Book: Publishing in Post-War America*. Chapel Hill: University of North Carolina Press.

Mazzolini, R. G., 1993. *Non-Verbal Communication in Science Prior to 1900*. Florence: Olschki.

Ottaviani, J., Myrick, L., 2011. *Feynman*. New York: First Second Books.

Peters, H. P., 2021. Scientists as Public Experts: Expectations and Responsibilities. In M. Bucchi, B. Trench (Eds.), *Routledge Handbook of Public Communication of Science and Technology*. London: Routledge, 114–128.

Riordan, M., Turney, J., (Eds.), 2000. *A Quark for Mister Mark: 101 Poems About Science*. London: Faber.

Rueda, A., Rosen, C., Crúz-Mena, J., 2024. "Let Science Be Told: A Review of Ideas for Storytelling in Science Communication". *Metode Science Studies Journal* 14, 151–157. https://doi.org/10.7203/metode.14.26522

Tatsuta, K., 2017. *Ichi-F: A Worker's Graphic Memoir of the Fukushima Nuclear Power Plant*. New York: Kodansha Comics.

Weitkamp, E., Almeida, C., 2022. *Science and Theatre: Communicating Science and Technology With Performing Arts*. Bingley: Emerald.

Wilson, E. O., 1998. *Consilience. The Unity of Knowledge*. New York: Alfred A. Knopf.

6

UNDERSTANDING DIFFERENT VIEWPOINTS ON SCIENCE COMMUNICATION
MORE COMPLEX, MORE PERPLEXING

A roundtable discussion at the 2023 conference of the PCST network turned on the question, where to start? The protagonists in the discussion were presented as if in opposition, one arguing that planning science communication starts with the audience, the other that it starts with the strategy of those initiating the action. It appeared like a complicated conundrum. But fairly quickly, it was evident that the audience or audiences for a proposed science communication action are a matter of choice, and that choice is based on a purpose. Equally, the development of a strategy is closely related to the selection of an audience or audiences; no one strategy can cover all the possible options. Many discussions in the field of science communication are characterised by oppositions of this kind, for example, between research and practice, professional and voluntary, science-qualified and not, perplexities that on further examination reveal themselves as dialectical complexities, where the two or more parts are inextricably and interdependently related.

In this chapter, we consider how the growth of the science communication field has brought with it complexities and competition. The field presents a somewhat contradictory appearance, spreading and consolidating, diversifying and even fragmenting. In the doing of science communication, the former is more the case, and in the thinking about science communication, perhaps more the latter. We ended Chapter 4 with reference to diverse historical and sociocultural perspectives. Here, we focus first on intellectual and theoretical perspectives.

DOI: 10.4324/9781032646749-6

There have been very few attempts to bring together the strands of science communication into a unified view. It is striking that two such attempts from the past decade point in very different directions. Davies and Horst (2016) present a view of science communication as cultural activity taking place in social structures defined by inequality of power and opportunity. In this perspective, analysing science communication is largely a matter of interpreting meanings, not just those intended by scientific interests but also those made by audiences of science communication. Jamieson, Kahan and Scheufele (2017) present a collection of essays by 57 authors, and thus not an entirely unified view, but held together by the notion that science communication is amenable to precise scientific analysis, notably in respect of public perceptions and reception of science communication, and that such analysis can – and should – direct a more effective practice.

Other handbooks aiming to summarise the whole field show yet other panoramas, or apply different lenses. Leßmöllmann, Dascal, and Gloning (2020) present a picture of the field that is informed mainly by theories and concepts from linguistics, rhetorics and philosophy. It is notable that contributors to this volume are comfortable referring to science communication research in these humanities contexts as the science of science communication, with no implication of the scientism underpinning the Oxford Handbook.

Our own handbook (Bucchi and Trench, 2008, 2014, 2021) contains chapters by authors principally rooted in sociology, mass communication and cultural studies. The trajectory of the book over three editions reflects the movements in the field; the title of the handbook from 2008 seemed a poor fit a decade later, given that contributors had been demonstrating the porosity of the supposed boundary between public and professional communication.

The difference between the last-named two handbooks is more complementarity than incompatibility. However, the contrast between the perspectives of the first two works cited above is a replay of the decades-old fault line between researchers favouring qualitative or quantitative analysis or, in the communication context, between 'administrative' research and critical research. Reviewing the Oxford Handbook, Mellor (2018) wrote that "the editors' insistence on a 'science of' science communication and their instrumentalist agenda to promote quantitative research aimed at improving future science communication coheres with the assumptions that science is capable of resolving social controversies and that communication is a problem that can be fixed — a communication fix to aid the technological fix that already seduces scientists and policy makers".

The quantitative and qualitative camps were also in evidence in the earliest days of the public understanding of science movement when public opinion surveys and narrative analysis, with case studies, were variously adopted as the preferred approaches. The administrative versus critical opposition manifested itself in communication research in the decades before that in a split among the German and Austrian communication researchers exiled in the United States. Whereas Paul Lazarsfeld, in founding the Bureau of Applied Social Research at Columbia University, proposed empirical, largely quantitative, social research around discrete social groups, his former colleague Theodor Adorno developed a cultural critique based on a view of the whole of society as unequal and communication as subject to manipulation.

In science communication, we also see competing theories of science-society and science-media relations, with resulting different definitions of the research needs. We find strong emphasis currently on scientific understanding of the public as the central concern of science communication research, as represented, for example, in studies of audience segmentation. The proposition for a science of science communication is focused on social-psychological studies of public (mis)perceptions. Closely related to this is the case made for evidence-based science communication and for science communication research to focus on building that evidence base that can be applied in improved practice, and the case to consider science communication as fundamentally strategic, developed with the aim to attain specific outcomes in targeted audiences.

The political allure of a theory of 'effective science communication' based on such conceptions is obvious. Philanthropic and other research funders in the United States, principally, came behind the science of science communication in the 2010s, which drew in researchers from political and risk communication with prior experience in the experimental psychological methods now finding favour in science communication.

The promise of evidence-based science communication is also evidently attractive. But the analogy with evidence-based medicine is not just imperfect – like all analogies – but misleading. Medical practice based on clinical trials of drugs giving detailed information on the effects of those drugs at specified dosages over specified time-periods is a very different proposition from science communication practice interpreting the results of disparate studies using diverse methods. It is not clear what evidence base science communication can refer to, and there is little agreement on how any such evidence base could be built.

There has been a noticeable trend in recent science communication literature for experimental research, which at least implicitly is aimed at providing evidence to guide practice. In such studies, different versions of a science communication stimulus – an article, or a video clip, for example – are given to selected groups in order to assess their responses. The question might be, how does changing the tone or the narrative approach affect their perception of the issue or their retention of information? Being typically based on small groups – frequently of students – the results are hard to generalise, and it is striking that very few of these studies have been replicated. The case for evidence-based science communication is closely related to that for considering it as always strategic: this approach posits that such communication starts with identification of specific, targeted audiences and that the selection of methods can be made with some precision on that basis. Common to these several positions is the assumption that science communication effects can be anticipated. But understanding communication effects beyond science communication remains one of the most problematic areas of communication studies. We shall return later in this chapter to the associated challenges of evaluating science communication activities and programmes.

RESEARCH-PRACTICE (DIS)CONNECTIONS

A proposition related to these views of science communication is that science communication research and practice are disconnected and could and should be more closely tied. This argument is among several that represent an exceptionalism commonly articulated in science communication circles: the claimed disconnect is stated as fact without context. We might reasonably ask, in what other domains is there a comparable relationship between researchers and practitioners? It seems safe to say that in journalism and politics, for example, research and practice barely acknowledge each other; in medicine, however, the relationship is well developed but highly mediated. It is somewhat ironic that the case about the science communication disconnect has been made repeatedly in contexts where researchers and practitioners coexist and cooperate, such as the conferences of the PCST Network. Many of those taking part in those conferences and in the network's leadership are themselves both researchers and practitioners.

JCOM – Journal of Science Communication, which routinely includes reflections on practice alongside research papers, published a collection of contributions that explored the relationship from a starting point that it needs to be fixed, though acknowledging that "bringing research and practice together is no easy task" (Fischer et al., 2024). The 72 abstracts received in response to the special issue editors' call for proposals indicated that the issues around this relationship are a live concern in the communities of research and practice. The published collection offered insights into successful collaborations and co-creations in support of the proposed reflection "on how research and practice might come together in a way that is mutually beneficial and enriching to both alike" (ibid.). This approach is a more nuanced version of a denunciation of the disconnection and a prescription to researchers that they ensure their work is more relevant to practice.

An alternative to the above set of related approaches to science communication research has questioned the possibility of producing evidence of the kind indicated above and considers the 'scientific' and strategic approaches as highly restrictive and reductive, both in terms of the problems posed and the perspectives brought to them. Cultural and STS approaches to science communication (e.g. Davies and Horst, 2016; Felt and Davies, 2020) see its practice as construction rather than – as in the dissemination model – translation. Science communication as a cultural activity creates new meanings in, through and around scientific ideas and information and a central task of science communication research is exploring those processes and those meanings; it is interpretive and narrative as well as descriptive and analytical.

At a more fundamental theoretical level, the relations between science, media and society have been addressed with the concept of mediatisation developed in Germany and Nordic countries (e.g. Weingart, 1998, 2012; Väliverronen, 2021). This offers new insights into the interactions of scientific systems and public communication systems. For those involved in its elaboration, "the emergence of the concept of mediatization is part of a paradigmatic shift within media and communication research" (Hepp, Hjarvard and Lundby, 2015, p. 315). Applied to science and media, the concept implies that media logics are (increasingly) applied within science, notably in preparing information for public consumption. It implies further that the public has become more meaningful to science, which seeks the public's attention through media, and that science's increased significance to the public is enhanced through media.

Scholars who have actively deployed the concept in their work have found, however, that it is not easy to operationalise in empirical research. Drawing on three data sets, Lehmkuhl, Promies, and Leidecker-Sandmann (2023) sought to demonstrate the working out of the mediatisation thesis but reported that neither does journalism do for public attention to science what the thesis suggests it might do, nor does media coverage influence scientific practice in the ways expected. This was not a falsification of mediatisation, they say, as the study focused on only some of the mechanisms at issue.

NAMES AND NEIGHBOURS

Among the issues giving rise to tension between the various perspectives outlined above is that of the scope of 'science' in science communication. Cultural and linguistic differences frame the approaches to that issue: science covers a wider range of knowledge production in, for example, French and German than it does in English. Even taking this into account, it remains the case generally that 'science' in science communication has referred almost exclusively to natural sciences. It is only latterly and hesitantly that communication of humanities and social sciences has been considered as part of science communication. In some of these disciplines, specific formulations and practices have been developed for public communication or participation, for example, in public history.

In our proposition of science communication as the social conversation around science, we seek to present an inclusive view of the research field as well as of the practice field. This view acknowledges that science communication is diversely defined as multi-, inter-, or transdisciplinary, though the differences in these definitions may be as much terminological as philosophical. Our own preference is to see science communication as interdisciplinary, defining this as a collaboration that integrates contributions from several disciplines, working together on shared problems. We argue that "at its interdisciplinary best, science communication is a continuing exercise in reflexivity on science and its place in wider intellectual and public culture" (Trench, 2023).

This element of reflexivity is crucial in the approach to and experience of science communication: scientists engaging in public communication can achieve heightened awareness of the purpose and potential of their work through conversations with those it affects; researchers in one branch of science can deepen their insight into it through engagement with those in other branches; active publics can strengthen their appreciation but also their critical awareness of how research is done through exposure to its workings. These various aspects of reflexivity are largely missing from the attempts to assert that science communication is a new (trans)discipline, defined within narrow parameters.

One way of broadening the field has been to adopt terms such as research communication or scholarly communication, which, in principle at least, can refer to studies of poetry as much as to studies of particles. In other related contexts, knowledge and knowledge systems are used as more inclusive terms than science. However, the stronger tendency has been towards fragmentation rather than consolidation. In some natural sciences–related fields such as risk, environment and climate, the choice has been to define new sub-sectors more or less independently of science communication. Some accounts of environmental communication present it as having no connection with science communication. This reflects at least in part the stronger advocacy element of communication around those themes but also the greater emphasis on engaging the emotions through imaginative and creative activity. The institutional settings offer incentives to fragmentation: conferences, networks, journals, chairs and departments can be designated by these newly declared specialisms; more individuals can be recognised as leaders and innovators in smaller niches.

A recent but more culturally limited branch from the tree is higher education communication, as discussed principally in German-speaking countries (*Hochschulkommunikation*). This brings science or research communication into closer connection with institutional communication, including marketing. The public engagement push by higher education and research institutions and the resulting central roles for full-time communicators formally educated in the field (see Entradas et al., 2024) contribute to the increasingly complex picture of science communication and associated fields. Institutional investment brings with it new professional opportunities but also vexed questions about the difficult reconciliation of a "promotional culture" (Väliverronen, 2021) with the more critical, reflexive studies of science communication in the same institutions. A 2023 PCST symposium on higher education and research institutions' involvement in science communication urged more resources but also more attention to the rationale and results of such programmes.[1] With the institutionalisation of science communication as a sector of the scientific infrastructure come increasing opportunities for commercial consultancies and training providers. The combination of institutional requirements and commercial ambitions contributes to an ideology of science communication as serving institutional purposes.

It also makes the organisation of science communication more challenging as sponsors seek to know the returns on their commitment; formal evaluation is required or recommended, though the methods for doing this remain contested, not least because the impacts or outcomes are difficult to anticipate and measure. The plurality of publics is highlighted here, for what is evaluated as appropriate for one audience segment may not be appropriate for another.

In the best of circumstances, evaluation of cultural, including communicative, activities is challenging. The experience of taking part in such activities may be diffuse and delayed in ways that make recounting it, never mind measuring it, very difficult. The highly variable conditions in which an individual or group may watch a television programme, visit an exhibition, or converse over a meal about a film they have recently seen mean that inputs and outputs are rarely clearly defined. In science communication, as in museum practices or media consumption, formal protocols have been advanced for evaluation before, during and after the relevant actions. At best, they may allow comparisons of reported satisfaction and stimulation between cohorts of participants, over time or across demographics. But the expectations of sponsors and organisers are generally set higher than that – they hope to see a return in changed attitudes or behaviours on the investment of human and material resources. This mismatch is the probable explanation for the frequently demonstrated discrepancy in science communication between the verbal commitment to evaluation and its actual implementation.

Evaluation undertaken with a view to improving quality has obvious value. In longer-term projects, it can be done *in itinere,* for example, through individual and group interviews of selected actors and audiences, and lessons drawn from these are applied in later versions. This is less formal and less well defined than the information that consultancies specialising in evaluation claim to produce, but it may well be more authentic.

From the perspective of science communication as the social conversation around science evaluation may be considered an effort to assess what any given action contributes to the conversation – does it halt it, widen it, divert it, amplify it, etc.? Such evaluation is more formative than summative, providing feedback from and to all participants within a continuing process rather than attempting a summary statement of outcomes. The kind of information gathered from closed-question surveys is less relevant here than that gathered from open-ended questions, whether written or spoken or both. A narrative account of comments, observations and judgements derived from such interactions hardly amounts to "metrics" but may be the most honest way of assessing the experience. The science communication communities have yet to come up with a widely accepted form of evaluation, and this may be the direction in which to look for it.

DIVIDED BY A COMMON LANGUAGE

Developments in cultural and communication studies and in STS continue to have much to contribute to the field of science communication research. Following the trends in these related fields, science communication research needs to give close attention to questions of language in two senses – the use of features such as metaphor to convey scientific information and ideas and the impacts of English-language domination in facilitating intercultural exchange but also limiting the exploration of differences. Some differences of view within the international science communication communities have more to do with various understandings of the apparently same words or various terms being used to refer to the apparently same phenomenon. It requires some familiarity with or competence in languages other than English to make sense of such differences.

Theoretical physicist and cultural commentator Jean-Marc Lévy-Leblond lamented the irresistible rise of English as the common language of science, saying it inhibited nuanced expression (1996, 2010). If that is true for science, it is true *a fortiori* for science communication. The enthusiastic adoption of English as the shared means of communication in a growing intercultural community has obvious benefits, and non-native speakers demonstrate over and over their adaptability in facilitating relatively seamless exchanges. But precisely because science communication is about meanings, it needs more refined means of expression than English as a second (or third) language can typically be. The heavy reliance on English as the common language of science and thus of science communication also tends to obscure the contributions to this field from outside the anglophone sphere.

The difficulties of using the phrase scientific (or science) culture become manifest when the anglophone and francophone spheres meet, as in bilingual Montréal in 1994 at the conference of the PCST network, hosted under the theme, When Science Becomes Culture. Lévy-Leblond was speaking about culture scientifique/scientific culture when a leading member of the British science communication community asked in a somewhat exasperated tone what was this thing he was calling scientific culture. Lévy-Leblond reminded him that the English writer CP Snow had spoken of scientific culture in his infamous Two Cultures lecture of 1959. Indeed, that lecture triggered a long-running and very English debate about the disconnection between the humanities (what Snow called "literary culture") and the sciences.

Snow has been somewhat unfairly saddled with the responsibility for inventing the binary and provoking the debate. He was conscious of the danger of his argument, noting that "the number 2 is a very dangerous number" (Snow, 1959/1993, p. 9). But all these years later, the concept of scientific culture remains a source of irritation in the science communication community, not least because it can mean at least two distinct things – the culture of scientific communities and the standing of science in public culture. Snow used these two meanings knowingly, remarking in a Second Look at the two cultures four years after his lecture that "it isn't often one gets a word which can be used in two senses, both of which are explicitly intended" (Snow, 1959/1993, p. 64). We might say that the two principal meanings are at the community and societal levels.

The in-built ambivalence of the term and the specific difficulty that native English speakers typically experience in using it have been amplified by the attempts to produce measurements of scientific culture through adaptation of traditional literacy and attitude surveys (see e.g. Godin and Gingras, 2000). Even accepting that not all linguistic communities will be able to adopt the term comfortably, it matters to know with some precision what its usages and connotations are. It also matters to acknowledge that it is frequently and comfortably used in Latin-based languages.

A possible definition of scientific culture at the societal level is that it refers to the presence and status of science in the general culture, as represented, for example, by the visibility of scientists, both historical and contemporary, and the attention to scientific ideas and processes in public discourse. Based on such a definition, scientific culture can be seen as the primary context within which science communication takes place.

Among other keywords presented later in our lexicon, (public) engagement with science is also the source of intercultural misunderstandings. Again, a specific event in Britain led to the wider use of the term. This was the House of Lord Select Committee Report on Science and Technology (2000) which famously observed "the mood for dialogue" around science. On the basis of that report, public engagement with science was proposed as a superior alternative to public understanding of science. Quite quickly, however, the term became muddied by inconsistent usage: it has been used to refer to both activities and effect. In the former usage, it is barely distinguishable from public communication or dialogue. The latter usage is more distinct but less common; it refers to deeper public interest and involvement in science achieved through public communication efforts. That is the meaning that travels more easily: a collection of essays by Latin American scholars was titled Science Communication and Public Engagement (Vasquez-Guevara et al., 2023), thus distinguishing the two dimensions of action and reaction. It is worth noting too that anglophone domination of the field obscures the use of engagement in French to refer to commitment, and often specifically political commitment. So, the writer *engagée* acts out of her commitment to a particular political worldview.

Automated translation can aid but also hamper intercultural communication around these and other terms in science communication. The Spanish *divulgación* and Portuguese *divulgação*, which are equivalent to dissemination, turn up in online translations as disclosure, which has a quite different meaning of revelation. Latin languages cannot easily distinguish criticism and critique, trust and confidence, distinctions that are germane to major issues in science communication. As for Lévy-Leblond's case for science's *mise-en-culture*, the English versions need a longer, clumsy phrase, roughly 'the placing of science into culture', which misses the original laboratory connotation – putting an organism into a (liquid) culture.

As we have seen, the international and intercultural spread of science communication brings complexities; it requires linguistic sensitivity to be celebrated, as it deserves to be.

HAVE WE EVER BEEN SATISFIED WITH SCIENCE COMMUNICATION?

The argument that we live in times of great change is a common thread in the reflection on science communication in most historical phases and contexts. The public understanding of science movement in the mid-1980s was triggered – among other things – by the perception within the scientific community that public attitudes towards science were becoming increasingly negative and an appropriate communicative response was required to counteract this trend. The later "new mood for dialogue" between science and society emerged in policy contexts, invoking a change and often a radical depart from top-down, paternalistic, one-way strategies of science communication that characterised the previous two decades.

More recently, the impact of the transformation in the technologies and forms of communication has often been considered a source of major transformation of science communication, outlining both new opportunities and substantial risks. A possible question then is: have there ever been periods of continuity in science communication in which actors and scholars did not have the perception of substantial transformations and required change? Or, more provocatively, have we ever been satisfied with science communication as it was? And if not, why so? After all, although we are far from consensus in defining quality in science communication, we often agree in nominating high-quality actors and products when looking back to the history of science communication: TV programmes, popular science books, magazines, public initiatives.

It is not possible to answer this question in a comprehensive manner here, but we can highlight some elements of continuity across the history of science communication based on elements already described in this book.

The centrality of communication in science: the emergence of modern science, as discussed in Chapter 2, was a radical transformation in the forms of creation, validation and dissemination of knowledge with communication at its core. Among modern science's significant elements of discontinuity was that knowledge was not valued until it was circulated and communicated.

The importance of communicating with non-experts: in the process just described, scientists discovered that allowing communicative exchange with outsiders was as important as establishing boundaries to protect their autonomy. As science became specialised and institutionalised, popularisation emerged as a distinct genre to make scientific content accessible to general, non-expert audiences.

Public talks and performances: public demonstrations and talks about science have been popular – with variations in format – across the centuries; public experiments in front of large citizen audiences were conducted by some of the greatest scientists in history.

From TED Talks to YouTube videos to spectacles on rock-concert scale, public performances by scientists, authors and science communicators remain a mainstay of science in culture.

The role of visible scientists: visible scientists have played a role in public communication of science through such public talks and demonstrations and popular science publications. From the mid-20th century, television contributed to making some of these known to mass audiences, including celebrities, whose private lives were material for public consumption.

Such elements of continuity are obviously not meant to understate relevant changes in the practice of science and of science communication, and in the broader policy, social, cultural, political, technological and media contexts of science. However, some of the drivers, dynamics and even formats of contemporary science communication can be seen to be in substantial continuity with longer-term traditions.

So why do the science communication communities invoke the need for change? A more critical view of this question can be related to a still unsettled recognition of science communication as an established professional and scholarly domain (Trench and Bucchi, 2010). There is still a widespread tendency – both among scientists and policy-makers – to ignore historical background, extensive and significant results, literature and professional expertise in our field in favour of stereotypes and prejudices about both audiences and communication processes. For example, it is widely taken as axiomatic in science circles that public audiences expect and require certainty and simplicity in statements from and about science. During the pandemic we could observe, on the other hand, that significant sectors of the general public appreciated hearing from scientists how and why they differ on given topics and what is not (yet) known with confidence. Also, institutional incentives operate to obscure the historical experience, encouraging innovation *per se* in science communication that often requires science communication to be formally experimental or even extravagant instead of consolidating and building upon existing practice and knowledge.

A more positive interpretation of the focus on change can be related to the intention to improve the quality of science communication. Writing in the scientific journal *Nature*, the great pioneer of science fiction, HG Wells (1894), criticised the popular science of his time with regard to style, clarity, lack of audience intelligence and humility. In an anticipatory criticism of the diffusionist deficit approach, he wrote:

> Out of a quite unwarrantable feeling of pity and condescension for the weak minds that have to wrestle with the elements of his thought, the scientific writer [forgets] that whatever status his special knowledge may give him in his subject, the subtlety of his humour is probably not greatly superior and may even be inferior to that of the average man, and that what he assumes as inferiority in his hearers or readers is simply the absence of what is, after all, his own intellectual parochialism. The villager thought of the tourist a fool because he did not know "Owd Smith". Occasionally scientific people are guilty of the same fallacy.
>
> (Wells, 1894, p. 50)

In many respects, the criticism by Wells (as well as the hoped-for improvement) still potentially applies to contemporary science communication. One possible and apparently paradoxical conclusion is that the focus on change is itself an element of continuity in the history of science communication.

ONCE MORE WITH FEELING: WHAT IS SCIENCE COMMUNICATION FOR?

Some years ago, with a large group of colleagues, we were involved in an extensive science communication training programme through the European Union-funded ESConet project. The training was aimed at researchers from all fields who wanted to know more about science communication and its wider social, cultural and policy contexts. The introductory module, "Who are you communicating with, and why?", triggered very interesting discussions about reasons and rationale for communicating science. Once participants had accepted to step out of the ideological tautology "science has to be communicated because it has to be communicated" (or because it is generically "a good thing"), various aims and motives started to emerge: strategic (raising the visibility profile of your institution, research group, or yourself; trying to support views or attitudes towards certain implications or applications of science); policy calls or duties (sharing knowledge outputs with citizens who have paid for your research with their taxes); personal (enjoying communicating science and hoping audiences will enjoy it too). It soon appeared clear to participants that they had a mix of reasons, often implicit or unclear; also, the resulting communicating efforts were often misaligned with such reasons.

Since then, just as actors and audiences of science communication have become more diverse and fragmented, motives and drivers have become more diverse and even contested. "Arms race for visibility or science substance?" is the question Bauer and Entradas (2022) posed – with a strong implication of the authors' answer – in relation to the massive growth of investments and activities in science communication by universities and research organisations during recent decades. Unclear aims fundamentally challenge functional models of science communication: if it's not clear why you are communicating science, how could you be able to eventually assess its impact?

Our vision of science communication as social conversation allows a plurality of aims by different actors and audiences, which need not necessarily converge. Examples of "peaceful divergence" are easy to find, for example, in a classic hands-on activity or exhibition organised by a department or institute: the organisers may see it as a way to present their research activities, secure coverage in the media, etc.; audiences may enjoy it as a way of spending leisure time; parents may use it as a way to keep their children happy. Who is right and who is wrong? All, and nobody. As conversation, science communication is inclusive of different motives and judgements. Also, compared to functional approaches, it is not necessarily limited in space and time as it deals with processes rather than timed communicative acts. Maybe some of those attending will not change their mind with regard to a certain topic. Maybe some will soon forget most of the content, as well as the names of researchers and institutions. But it could well be that after the event, one or more participants will start to follow the institute or the scientist on social media; or one day, read a book about that topic; mention it to somebody in another conversation. When selecting a university course or advising someone on what topic, they may recall that enjoyable event.

NOTE

1. See symposium closing statement at https://www.univiu.org/focus-areas/science-communication-and-education/2023-pcst-venice-symposium

REFERENCES

Bauer, M., Entradas, M., (Eds.), 2022. *Public Communication of Research Universities: Arms Race for Visibility or Science Substance?*. London and New York: Rourtledge.

Bucchi, M., Trench, B., (Eds.) 2021/2014/2008. *Routledge Handbook of Public Communication of Science and Technology*. 3 edn. London and New York: Routledge.

Davies, S. R., Horst, M., 2016. *Science Communication: Culture, Identity and Citizenship*. New York: Palgrave Macmillan.

Entradas, M., Bauer, M.W., Marcinkowski, F., Pellegrini, G., 2024. "The Communication Function of Universities: Is There a Place for Science Communication?. *Minerva* 62, 25–47. https://doi.org/10.1007/s11024-023-09499-8

Felt, U., Davies, S. R., (Eds.), 2020. *Exploring Science Communication: A Science and Technology Studies Approach*. London: Sage.

Fischer, L., Barata, G., Scheu, A. M., Ziegler, R. 2024. "Connecting Science Communication Research and Practice: Challenges and Ways Forward". *JCOM* 23 (02), E. https://doi.org/10.22323/2.23020501

Godin, B., Gingras, Y., 2000. "What Is Scientific and Technological Culture and How Is It Measured: A Multi-Dimensional Model". *Public Understanding of Science* 27 (1), 47–58.

Hepp, A., Hjarvard, S., Lundby, K., 2015. "Mediatization: Theorizing the Interplay between Media, Culture and Society". *Media Culture and Society* 37 (2), 314–324.

House of Lords Select Committee Report on Science and Technology, 2000. *Science and Society*. London: The Stationery Office.

Jamieson, K. H., Kahan, D., Scheufele, D.A., (Eds.), 2017. *The Oxford Handbook of the Science of Science Communication*. Oxford: Oxford University Press.

Lehmkuhl, M., Promies, N., Leidecker-Sandmann, M., 2023. Repercussions of Media Coverage on Science? A Critical Assessment of a Popular Thesis. In I. Broer (Ed.), *The Science-Media Interface – On the Relation between Internal and External Science Communication*. Berlin: De Gruyter, 139–160.

Leßmöllmann, A., Dascal, M., Gloning, T., (Eds.), 2020. *Science Communication*. Berlin: De Gruyter.

Lévy-Leblond, J. M., 1996. La langue tire la science. In J.-M. Lévy-Leblond (Ed.), *la Pierre de Touche: la science à l'épreuve*. Paris: Folio Essais, Gallimard, 228–251.

Lévy-Leblond, J. M., 2010. "Quand la langue tire la science". *L'Archicube* 9, December, 52–61.

Mellor, F., 2018 "Book review: Jamieson, K. H., Kahan, D., Scheufele, D.A. (Eds.), 2017. The Oxford Handbook of the Science of Science Communication". *Public Understanding of Science* 27 (6), 750–754.

Snow, C. P., 1993/1959. *The Two Cultures*. Cambridge, UK: Cambridge University Press.

Trench, B. 2023. "Strengthening Interdisciplinarity in Science Communication Education: Promise, Pleasures and Problems". *JCOM – Journal of Science Communication* 22 (6), Y02. https://doi.org/10.22323/2.22060402

Trench, B., Bucchi, M., 2010. "Science Communication, an Emerging Discipline". *JCOM - Journal of Science Communication* 9 (3), C03. https://doi.org/10.22323/2.0903030

Väliverronen, E., 2021. Mediatisation of Science and the Rise of Promotional Culture. In M. Bucchi, B. Trench (Eds.), *Routledge Handbook of Public Communication of Science and Technology*, 3rd edn. London and New York: Routledge, 129–146

Vasquez-Guevara, D., McIntosh White, J., Weiss, D., Ochoa-Aviles, A., Ortiz-Vizuete, F., (Eds.), 2023. *Science Communication and Public Engagement: Evolving toward Science-Society Participation*. Lanham, Maryland: Lexington Books.

Weingart, P., 1998. "Science and the media". *Research Policy* 27 (8), 869–879.

Weingart, P., 2012. The Lure of the Mass Media and Its Repercussions on Science. In P. Weingart, M. Franzen, S. Rödder (Eds.), *The Sciences' Media Connection – Public Communication and Its Repercussions: Sociology of the Sciences Yearbook*, 28. Dordrecht: Springer. 17–32.

Wells, H. G., 1894. "Popularising Science". *Nature* 50 (1291), 300–301.

7

CHALLENGES AND PRIORITIES FOR SCIENCE COMMUNICATION

A theatre in the centre of Paris, 1881

Good evening, welcome and welcome. This evening we have with us a great character, truly one of those figures that the whole world envies us, Professor Louis Pasteur. Chemist, microbiologist, member of the Academy of Sciences, publications, awards, truly, believe me, an impressive CV. In recent years, thanks to his successes with vaccination, we have all come to know him and his beautiful white beard. Come on, let's give him a big round of applause!

So, Professor, our theme this evening is education and merit. How fair is it to reward merit in our school?

(Pasteur, embarrassed) And what do I know? I deal with animal diseases, anthrax, chicken cholera. In this regard I am carrying out some very interesting experiments...

Okay, professor, can you tell us about these another time, eh? So, merit and education. We've been talking about it for days, you must have formed an opinion on the matter, right? Are you with the minister or against the minister?

But I don't know, look, I haven't really thought about it enough...

(Host now impatient) Okay, professor, but sorry, when the editorial team contacted you you agreed to participate...

Yes, but they told me I could say anything I wanted, and I thought I would talk about my studies, the importance they have for breeding and nutrition...

Of course you can say what you want but about the topic of the show, not about what you like!

If you want, I can talk about my studies on silkworm diseases...

DOI: 10.4324/9781032646749-7

But who wants to be interested?

Chicken cholera?

Yes, so the audience who has just finished dinner gets up and leaves disgusted...

I could talk about rabid disease.

Here you are! Let's talk about rabid kids when the competition for merit is too strong...

Actually, I was referring to the disease that is transmitted through the saliva of animals...

Enough professor, I give up. Let's change subject. Do you want to talk about excise taxes on horse-drawn carriages? Is it right for the Government to intervene to sterilize them?

The horses?

(discouraged, snorting) But no, the excise duties...

(conciliatory) If you want a more accessible topic, I could talk about my research on beer fermentation.

But no, we can't advertise alcohol here this time of the day... listen, if you're strong on gastronomy, why don't you talk about the methods of cooking pasta?

(startled) Some pasta? But I don't know, I don't think I've ever eaten it...

Oh hell, professor, a minimum of flexibility, you are now a recognizable face, a public figure... okay, the time available to us is up for this evening. See you here tomorrow evening to talk about the death penalty and life imprisonment with Dr. Annelise Persil, professor of comparative zoology at the University of Toulouse, who at least always answers everything and isn't as difficult as this complaint from Professor Pasteur. Good night.

Louis Pasteur may never have faced such a difficult encounter with the public media, but as a proto–celebrity, he had to negotiate the challenges and risks, as well as the opportunities in public engagement and visibility. In this chapter, we explore current challenges and opportunities for science communication, and we make some proposals on how it might face these. One central consideration is that science communication is increasingly filtering through the arts and popular culture. While dedicated spaces for science communication still exist, limited by the self-selection of participants and audiences, science content increasingly inspires, and transpires into, other contexts not explicitly labelled as science communication: films, artworks, pop songs, novels and comics.

These platforms are not oriented to purposes such as the traditional focus of science communication implied. They are not designed to transfer science content or to support positive attitudes to science and its applications. Such contexts are in principle oriented toward pleasure, entertainment, triggering emotions among viewers, readers or listeners. Science may be becoming an inspirational refuge and reservoir of new ideas for many cultural forms.

A distinction, however, is needed, particularly in view of recent developments. Traditionally, films, novels or pop music are produced and released under a commercial model, supported by paying viewers, readers or listeners. The producers are not committed to endorsing a certain image of science or of specific science fields. They use science as a source of inspiration, drawing suggestions, impressions, pieces of content and imagery from science. They may value credibility more than accuracy in the strict sense. Audiences, for their part, bring in their own perceptions, expectations and emotions. The outcomes of this process are unpredictable and can take multiple forms: commercial success, critical appreciation, contestation, critique and further inspiration. There have been cases of fictional work having a significant impact on the public perception of social issues related to science: films like *The China Syndrome* (1979) or *Philadelphia* (1993) contributed to raising social sensitivity for the potential risks of nuclear power and AIDS disease.

However, a new production model has emerged in connection with the increasing science communication efforts by research and academic organisations. In some cases, these organisations support the production of fictional works and artworks connected with their areas of activity. The spectrum of such support is broad and goes from hosting fellowships or residencies to covering professional and production fees. Similarly, the mandate can be more or less specific, leaving it to artists to decide about content or more strategic and specific aims. National and international research organisations like CERN, NASA or ESA have commissioned ballets, music compositions and theatre pieces; they have hosted artists and their works. Foundations and founding agencies have supported artists to work on science topics.

FROM VISIBLE SCIENTISTS TO CELEBRITY SCIENTISTS

With the increasing popularity of the new visible scientists, many acquiring celebrity status, their features and roles in the media and in the public sphere more broadly are changing. Just as the rise of science as a source of creative inspiration may reflect a crisis of certain cultural genres, likewise the rise of visible scientists and celebrity scientists could be linked to a loss of credibility of and interest in political figures. This hypothesis finds confirmation in empirical data from international studies of public perception of science, where scientists feature highly in terms of public trust not only in absolute terms but also in relative terms, that is, compared with politicians or business leaders.

Another driver of the personal visibility of scientists may connect to changes in the media landscape. Social media, in particular, value personal visibility, authenticity and recognisability. One of the possible ways to summarise trends in media communication could be that people struggle to become brands, while brands and organisations struggle to become more personal and familiar. Some faces of scientists (including scientists from the past) have become commodities, communicative currency that can be circulated on platforms through clips, memes and quotations, their use largely extended beyond their area of research and expertise. The most obvious example is that of Albert Einstein, whose strikingly recognisable face and whose pithy quotes turn up in many contexts.

We should not see this as a completely new trend but rather as amplifying and transforming a tendency that goes back at least to the early 20th century. When science was gradually becoming a more collective, organised enterprise, it needed ways to communicate in a more personal, incarnate, human form: stories of human beings who had achieved great things. Part of the key to the extraordinary success of the Nobel Prize, established in 1901 on the basis of the will of scientist and entrepreneur Alfred Nobel, was that it came at the right time in history, offering society faces and stories to materialise and visualise a science that was becoming a less personal and individual enterprise (Bucchi, 2018, 2025).

Personal visibility and celebrity have implications for science communication, some of which still need to be explored and understood. While trust in science, scientists and science organisations remains very high in most countries, evaluation of communicative performance by individual scientists is much less positive, being perceived as a source of potential confusion rather than clarification. Personalisation lends itself to the representation of contrasting positions, particularly under certain formats (e.g. television talk shows). As individuals, scientists can take sides or positions, including political positions. During the COVID-19 pandemic, in several countries, including Sweden, Brazil and the United States, some visible scientists became controversial in their advisory role to politics and, in some cases, came into explicit conflict with politicians (Joubert et al., 2023). In Italy, some scientists who had become famous during the pandemic ran for parliament, which led to discussions about their previous independence as advisors or commentators.

Other consequences can be seen, as mentioned, in tensions between communicative aims, strategies and identities by individual scientists and the organisations with which they are affiliated. Organisations can respond to this tension in different ways, more softly (e.g. expanding and intensifying their communication support with more staff or training courses for scientists) or more sharply (trying to contain or even suppress individual communication activism, something not easily done in most democratic, constitutional contexts).

THE QUALITY CHALLENGE

The quality of science communication remains a key challenge in different ways. How do we recognise quality in science communication? It should have become clear from the previous chapters that, particularly in a framework of science communication as social conversation, quality has multiple dimensions beyond the traditional meaning of "accuracy" that was emphasised by diffusionist views and models, where "good science communication" was science communication that "accurately transported" (whatever that might mean) a piece of scientific content from specialist production down the alleys of popular communication.

That vision was problematic, and it is even more problematic today. The prestige of the source or channel cannot be used anymore as a proxy for the quality of content.

Social media host myriad quotations dubiously attributed to scientists as a means of gaining authority and prestige. The magazine *New Scientist* collected a long series of quotes attributed to Einstein, including one widely spread on the disappearance of bees, that were never actually said or written by the famous physicist. "A scientist said it" is increasingly and confusingly used as a synonym for "scientific".

The quality of public communication of science is—even more than in the past—highly dependent on the quality of research produced and published in specialised contexts. In the context that we have described elsewhere as a "crisis of mediators", new research is increasingly pushed in real time into the public domain without being "filtered", as was the case in past decades, by professional mediators and popularisers. This inevitably connects science communication at large with trends causing major concerns in the world of research policy and academic publishing: a significant rise in the number of retracted papers from 30 cases in 2002 to more than 600 in 2016, in the Medline database alone; the emergence of "predatory journals" offering to publish any content regardless of its quality; and a lack of replicating studies and experiments. The now fully discredited study on the link between vaccines and autism was at the time published by the prestigious medical journal *The Lancet*; the same holds for other studies later proven to be false (or even fraudulent) after their appearance in significant journals.

Some cases bear particular interest for the field of science communication. In 2016, the journal *Science* published a paper by scientists from the University of Uppsala, Sweden, according to which exposure to high concentrations of polystyrene would lead to some fish larvae "preferring to eat plastic rather than their natural prey". The paper's conclusions obviously appealed to multiple media frames, and they made headlines globally. "Fish eat plastic like teens eat fast food, researchers say", summarised BBC News. The paper was retracted by the journal in May 2017 following accusations of data fabrication. However, further reports revealed that the journal had earlier dismissed strong criticism of the paper and its empirical basis submitted by a non-academic, amateur scientist member of the American Association for the Advancement of Science. This led a science journalist to raise the questions: "Does citizen science count for nothing in academia? Are amateur scientists expected only to unquestioningly applaud and assist their academic role models, while keeping their scientific criticisms to themselves?" (Schneider, 2017).

We outline below other possible dimensions of science communication in contemporary society beyond accuracy, as represented in our AEIOU manifesto. Quality is not an abstract issue for theoretical discussions among science communication scholars. Instead, it has practical implications, for example, in terms of recognising and rewarding good quality science communication, particularly within research institutions. One of the many paradoxes of contemporary science communication is that institutions have a duty to engage with public communication with non-specialist audiences, but in most cases, they cannot oblige researchers to take part in such activities as part of their duties. In most contexts, such institutional work is documented through general reports describing the activities carried out for funding agencies, mostly without looking at their quality.

Clearly, assessing the quality of such activities is not straightforward. One possibility would be to develop a set of indicators, for example, on the basis of the dimensions outlined above and other possible dimensions that could be specific to different topics and activities. Another option would be peer review, which would imply a perhaps not-obvious definition of who are the peers – science communication practitioners, communication officers or science communication scholars?

A framework of science communication as social conversation around science expands and articulates the notion of quality, and not just in terms of multiple dimensions. As an interactive process, science communication cannot be judged only by one of the parties or actors involved (e.g. science organisations, scientists). The common concern with effectiveness obscures the quality issue: it is possible that a certain kind of authoritarian or populist communication may be judged effective in influencing attitudes or even behaviour. But does it meet acceptable ethical and aesthetic standards? Can it be considered of good quality?

In our conversation framework, quality could be seen as the potential to further stimulate and promote social conversations, which also places evaluation in a long-term, rather than short-term, perspective. Evaluation is more than answers to set questions; it is the basis of continuous quality improvement. Implementing such evaluation may not be straightforward: how, for example, do we assess the quality of a movie inspired by science? Is it on the basis of its commercial results, its critical reception, or its capacity to inspire further works or public discussions?

BEFORE AND AFTER THE PANDEMIC

The pandemic highlighted and amplified most of these challenges, representing a spectacular and unprecedented global experiment in science communication. Science had to be communicated in its making. Attitudes to science and uses of science communication challenged in many ways the received stereotypes. Data showed a return to traditional formats and outlets (news media, particularly television), scepticism about social media, trust in science and science institutions and criticism of the communication actions of some scientists. The pandemic crisis also highlighted the increasing communicative demands on researchers and their institutions, with many of them not being adequately prepared. It also pointed to communicative tensions between individual visible scientists and institutional communication and attempts to resolve them (Bucchi and Schäfer, 2025).

Longer-existing questions continue to have relevance and resonance for science communication; for example, the question, what is the science in science communication? This has come up in two slightly different contexts in our earlier discussions. It endures because the understandings of science continue to shift beyond their application in science communication. Another enduring question – how much scientific knowledge is needed to be a science communicator? – arises frequently, sometimes in tendentious form as a kind of provocation from members of scientific communities to the more recently established communities of science communicators. The related question, is what does it mean to be a science communicator today? relates to the scope of "science", as mentioned earlier, but also to the scope of the communication work. Is the science communicator someone who deals exclusively with news and information or also with broader, general issues such as downstream social implications? Are the functions of the scientist who is an occasional science communicator distinct and separate from those of the full-time professional, and in what way?

If there is a pattern showing here of placing question upon question, it is because we have to face the possibility that some of the issues mentioned here cannot be settled definitively. Earlier (chapter 6), we asked, have we ever been satisfied with science communication? In view of the ubiquity of science communication we ask further, if science communication is everywhere, do we still need to expand it or should our efforts be concentrated on understanding it better and improving it? Rethinking science communication has been the order of the day over recent years, as expressed in the EU project, ReThink, which looked in particular at the challenges and opportunities of the digital age but was almost immediately on its completion overtaken by the new digital challenge of artificial intelligence. Within weeks of the emergence of AI-driven ChatGPT in November 2022, its application to science communication was being demonstrated. If there is no financial return on science communication and it remains free of charge, could it be a candidate for partial automation?

GUIDANCE INSTRUMENTS

For orientation of future science communication we present a framework of the Modes of the Social Conversation around Science, revised from its earlier presentation, we propose a Pocket Manifesto for future science communication defined by AEIOU standards and we present a summary Research Priorities agenda.

The framework is intended to offer guidance in the choice of science communication approaches through making their underlying assumptions and implications more explicit. We consider modes of the conversation and models of communication to be near-equivalent; they refer to the assumptions underlying a chosen communicative action. Figure 7.1 is proposed as an aid to setting up, joining, or making sense of conversations around science. The spectrum illustrated in this way may be compressed or extended, like an accordion, in any period or over time. Distinctions like purposive and non-purposive should not be seen as a binary on–off but rather a greater or lesser emphasis on stated or unstated purposes of a communication. New formats of science communication, notably art–science projects, may well facilitate conversations of kinds not yet envisaged. The range of modes is continually growing, but not just in the direction of more participation or co-creation, as a "progressivist" point of view might suggest.

We can imagine that practices and platforms that we have reviewed earlier (Chapter 5) and, indeed, others can be overlaid on this spectrum as in-person platforms (from left to right):

public lectures; demonstrations; lab visits; exhibitions; public interviews; interactive science centres; science festivals; three-minute talks; science theatre; science cafés; science comedy; latest-generation science centres; participatory field trips; art-science co-creation; citizen science

and mediated platforms (left to right)

textbooks; popular science books; science fiction; science in literature; popular science magazines; science in current affairs magazines; newspaper science sections; TV documentaries; science radio programmes; science in general news; online videos; podcasts; visual communication; science in general interest radio and TV; social media.

It should be noted, in further emphasising that the range of models is not to be conceived as an evolutionary line, that practices and platforms to the left are as much in evidence in contemporary contexts as are those to the right.

Three concepts are included here that have been little or not at all discussed elsewhere in this book – technocracy, civic science and science critique. From a technocratic perspective, those with expertise in technical disciplines, including economics and management, as well as science- or engineering-based, are or should be in key

Modes of the social conversation around science

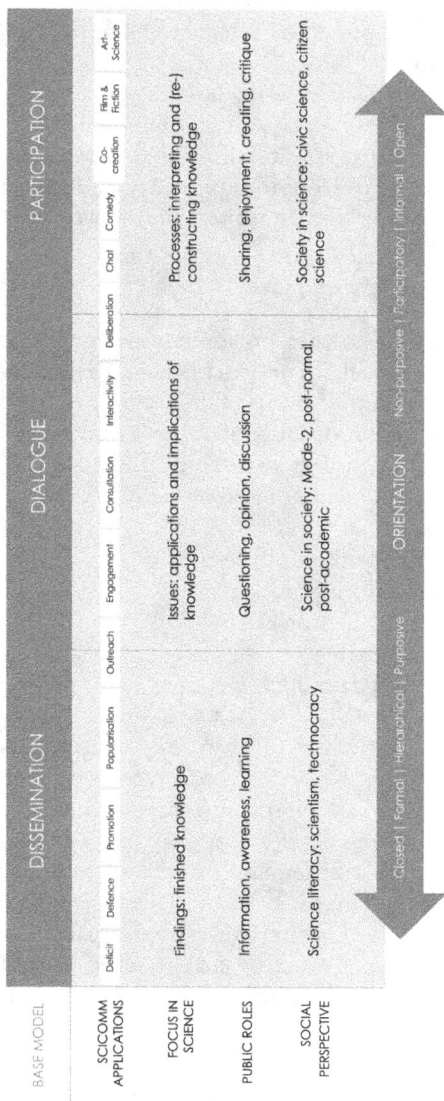

BASF MODEL	DISSEMINATION					DIALOGUE						PARTICIPATION				
SCICOMM. APPLICATIONS	Deficit	Defence	Promotion	Popularisation	Outreach	Engagement	Consultation	Interactivity	Deliberation	Chat	Comedy	Co-creation	Film & Fiction	Art-Science
FOCUS IN SCIENCE	Findings: finished knowledge					Issues: applications and implications of knowledge					Processes: interpreting and (re-) constructing knowledge				
PUBLIC ROLES	Information, awareness, learning					Questioning, opinion, discussion					Sharing, enjoyment, creating, critique				
SOCIAL PERSPECTIVE	Science literacy; scientism, technocracy					Science in society: Mode-2, post-normal, post-academic					Society in science: civic science, citizen science				

ORIENTATION

Closed | Formal | Hierarchical | Purposive ——— Non-purposive | Participatory | Informal | Open

Figure 7.1 Modes of the social conversation around Science. Based on Bucchi and Trench, 2021, adapted from Trench, 2008.

leadership roles. Civic science has been proposed at various times as an orientation for science to be fully embedded in the community or country of which it is a part, socially responsible and responsive. This has also been presented (Harney, 2003, p. 3) as science being "engaged and invited into the national dialogue" and "being worthy of the public trust". The case made for science critique or criticism also relates to science's social and cultural embeddedness; it is taken for granted that theatre, music, cuisine and fashion are reviewed critically as part of their belonging to society, and thus, it is argued, so should science. Some science writers fulfil that role.

POCKET MANIFESTO

The Pocket Manifesto presents aspirations and criteria for science communication, mixing the normative and descriptive. The AEIOU mnemonic based on the vowels of the English alphabet is knowingly adopted from a definition of contemporary science communication that applied it, though very differently (Burns et al., 2003). Science communication can be, should be or likely will be

- Authentic and aesthetically aware
- Engaged and ethically aware
- Interpretive and inclusive
- Open and original
- Uncertain and unfinished

Authenticity is particularly relevant in contemporary social media. "Good" communicative performances are such insofar as they are perceived as authentic, with scientists and science influencers bringing in their own personal life and character and not just their expertise (Geipel, 2020). It remains good advice to scientists entering the public media sphere to "be yourself" rather than to adopt a performative persona that is conceived as appropriate. In the COVID-19 pandemic, many experts had to engage with media from home, in family contexts, and without actual or figurative cosmetic make-up, and the evidence suggests that this authentic presentation enhanced their credibility.

Aesthetic awareness, like aesthetic quality, is difficult to define and assess. "Beauty" has guided scientists in certain fields to judge their own hypotheses and outcomes. "Our structure is very beautiful", wrote Francis Crick to his son about the structure of DNA he had devised with colleague James Watson. Elegance is a high value, particularly in mathematics and theoretical physics. Author Philip Ball tells the history of science through *Beautiful Experiments* (2023); his interest in aesthetics, including the aesthetics of science, has also given rise to books on music and colour. Ball writes vividly and elegantly on these matters, demonstrating how being aesthetically aware in science communication means giving attention to style. This concept has been proposed as a non-normative guidance for quality, bridging different areas of social life such as science, art, culture and craft-making (Bucchi, 2013). Eloquence in speech, aptness of analogies in writing and care in construction are all aesthetic qualities that are relevant to science communication.

Engaged science communication is committed to helping scientists be their best selves in public communication and helping individuals and groups in wider society to make sense of science, all as part of good citizenship. To be engaged requires communicators to develop and constantly renew their awareness of audiences, such that evaluation of their efforts is suitably attuned to different categories of participants – not just those defined as audiences, but also scientists and communicators taking – and using evaluation results to improve future activities.

Ethical awareness means taking care of the potential implications and consequences of science communication. A potentially technically accurate statement, "there is no evidence of a forthcoming earthquake, go back to your house and drink a glass of wine", as suggested by a scientist member of the Italian Government Risk Commission the day before the disastrous L'Aquila earthquake in 2011, or "we cannot tell at this stage whether Covid-19 vaccines will actually work", as stated by an immunologist on television upon the initial vaccine release, can have very dangerous impacts on the health and life of people. Communicating responsibly means taking account of audience and contexts; the same content or statement can have totally different implications in a specialist meeting and in a prime-time TV show. Too frequently, this is not well understood, and scientists resort to blaming the public, media and society for the supposed increase in distrust of science. The ethically aware science communicator is conscious both of their obligation to be honest and respectful in their communication and to ensure that ethical issues that may arise in and from the science they are communicating are adequately treated. On both counts, this may mean saying more about the conditionality of scientific findings than the institutions wish to. As science communication becomes professionalised, the autonomy of communicators to act as their conscience and standards indicate needs to be guarded.

Interpretation: Along with the exposition of science, which is the traditional content of science communication, comes its contextualisation and thus its interpretation. Science communication can indicate what this finding or this new direction in research means in broader scientific and social contexts. If science communication as the social conversation around science is meaning-making, then those meanings need to be laid out and considered anew in the continuing conversation. Science journalism is no longer the indispensable intermediary between science and the public; among its newer roles are those of curation of information from diverse sources (Fahy and Nisbet, 2011) and interpretation of scientific dissonance when it arises.

Inclusivity: "Science for all" is a traditional slogan of science festivals and other events, but science communication has more typically been for selected audiences, leaving others largely out of the picture. The current concerns about inclusion in social and cultural organisations and activities more generally have been applied in science communication mainly in relation to audiences. But inclusivity extends also to the scope of science and to the range of formats in science communication. In English usage, we have to make explicit the possible inclusion of humanities and social sciences in science communication, which may be implied in other Germanic or Latin-derived languages. Reflecting on science communication in humanities and social sciences contexts remains a recurrent task. As we have set out in different ways elsewhere in this book, approaching science communication as the social conversation around science is a more fully inclusive way of looking at the field.

Openness: In an age when open science is advocated even in major institutions, more open science communication seems an obvious need. In open science, the datasets, protocols and grounds of analysis are made explicit, and the source materials are accessible. By analogy, open science communication would lay bare its workings – this is who we are, this is how we know what we say we know and these are the limits of our knowledge. As mentioned frequently, there is much to be learnt from the COVID-19 pandemic, and one of the lessons is that saying publicly, We don't know this (yet), contributes to credibility rather than taking from it, as might have been feared. Openness and transparency in science communication are increasingly recognised as important bases of public trust in science.

Originality: This is a well-known value in science where the review of papers submitted for publication includes novelty as a criterion but is balanced with the need to see previously published results validated through replication. A similar balance needs to be struck in science communication, where the pursuit of originality for its own sake or for attention can be counter-productive. In the diversification of science communication formats, the capacity for invention and innovation has been demonstrated repeatedly and fruitfully. In more conversational formats, such as science cafés or science comedy, a capacity for improvisation is required. Fully scripted public communication is unlikely to be fully engaging. Going into it with an open mind and open ears means being ready to create something in the moment.

Uncertainty: there are two main dimensions to this in science communication: one is the need to give account of the uncertainty of scientific findings, and the other is to recognise that the outcomes of any communication are uncertain. These two combined are central challenges of science communication, the source both of frustration and of stimulation. At latest, through the COVID-19 pandemic, it came to be widely recognised that science communication cannot and should not deal only with certain, several-times-validated, scientific knowledge. Science proceeds by reducing uncertainty; findings are much more often assessed in terms of degrees of confidence than as new monuments of knowledge. Science communication is increasingly called on to give account of the process of producing knowledge, with its attendant uncertainties. As science communication does this and so many other things, it has to face the uncertainty of its own reality: for most communication, in most circumstances, there is no guarantee of achieving predetermined effects.

Unfinished: there is a before and after to all science and to all science communication: no experiment and no project starts from a blank sheet and none closes the book definitively. A book, a show, or an event tunes into a pre-existing conversation, enters it and leaves it with new elements added but also new insights gained. Doing this respectfully is a mark of good science communication; it is also likely to contribute to learning that supports further improvement. The work of science communication is never done.

RESEARCH PRIORITIES

We present our outline view of current and prospective research around four foci, Publics, Actors, Representations and Contexts. Many individual pieces will cross over two or more but, by their own design – as reflected, for example, in the authors' chosen keywords – rest primarily in one.

Publics

Here the research focus is on audience and participant groups; studies may consider, among other things, their demographic make-up, cultural and educational background, ideological or other beliefs, attitudes to science and other instances of authority, their reception and sense-making of science communication experiences and their level of interest in being more active participants in such processes. This has been the dominant focus in science communication research in recent years, and this trend prompts us to two cautions: research should be careful 1. to avoid making judgements on publics, either that they are insufficiently aware, engaged or active in science – publics are entitled not to know and not to be involved; and 2. to avoid assuming that science communication experiences are solely about adding to knowledge – they may have affective as well as cognitive consequences, for example, stimulating pleasure through the act of taking part.

Actors

Scientists and communicators, their interactions and their respective motivations were more extensively studied in the earlier days of science communication research. Scientific institutions have been getting more attention recently, as more of these become active sponsors of science communication activities; the inducements and the obstacles they present to their researchers who pursue public engagement are a particular object of interest. But more attention needs to be paid to the deeper culture of scientific institutions as it influences scientists' expectations and beliefs; this may be a more important dimension of researchers' participation – or not – in public communication than specific rules or policies. The actors of science communication have become more diverse, extending into what might have been defined earlier as the publics, through non-governmental organisations, citizen science and public patient involvement. The roles of designers, photographers, curators and science centre staff need more analysis.

Representations

Research with this focus looks at where, how much and in what forms science is held up to public view in mass media and other outlets. Studies of media coverage of science tended strongly for some decades towards elite newspapers and television documentary formats with their specialised science coverage. Popular newspapers and general-interest magazines and radio featured much less, if at all. The availability online of text databases reinforced for some time the disproportionate attention to up-market newspapers, and these were widely analysed in terms of how they framed science or particular science-related issues. However, the later development of social media changed the game both because these platforms facilitated unmediated communication by researchers and their institutions with various publics and because members of those publics became protagonists in the communication through comments, feedback and other user-generated content. Studies of science in social media may be focused on publics, on scientific actors, or on ways of framing issues. Social media may also be a means of studying social representations or "imaginaries" of science and scientists within communities.

The ways in which films represent science and scientists have been the object of important studies, but such representations in other audio-visual media such as television current affairs and entertainment programmes or radio in general and talk-back radio in particular have received much less attention.

Contexts

In referring above to institutional cultures and the socialisation of scientists within them, we were touching on one of the contexts relevant to understanding science communication processes. This institutional and professional context is the source of the ideological framework of scientism, which we discussed earlier and which has hardly been considered in science communication research. The wider context of a country's political and civic culture is important here too, not just in shaping institutions and other formal organisations but also in influencing how citizens regard the state, government and other authorities and elites. As political contexts change, they also influence how groups and individuals balance social and individual responsibility with gain, including how science acts with and for society.

We observe that the volume of recent and current published work in the Publics category exceeds that in the other three categories combined and believe the balance needs to be redressed and more attention given to the topics that we have indicated above under Actors, Representations and Contexts.

WHAT'S IT ALL ABOUT? WHY DOES IT MATTER?

We conclude by asking again and aiming to answer, What is science communication all about? And why does it matter?

Science communication is the social conversation around science through which information, knowledge and ideas from and about science are shared. It facilitates connections and interactions between communities involved in, engaged with or affected by scientific knowledge. It is the means by which the workings of science are made accessible and science is situated within culture.

Science communication matters because science matters well beyond the scientific communities where it is practiced. Science has generated knowledge that has shaped history and influences the daily lives of all in society. Science communication makes visible how this happens.

REFERENCES

Ball, P., 2023. *Beautiful Experiments: An Illustrated History of Experimental Science.* Chicago: University of Chicago Press.

Bucchi, M., 2013. "Style in Science Communication". *Public Understanding of Science* 22 (8), 904–915. https://doi.org/10.1177/0963662513498202

Bucchi, M., 2018. *Come vincere un Nobel. Il premio più famoso della scienza.* Turin: Einaudi.

Bucchi, M., 2025. *Geniuses, Heroes, and Saints. The Nobel Prize and the Public Image of Science.* Cambridge and London: The MIT Press.

Bucchi, M., Schäfer, M. S., 2025, "Who Speaks for Whom? Tensions between scientists and their organizations in the public communication of science: sources, dimensions and ways forward".

Bucchi, M., Trench, B., 2021. *Routledge Handbook of Public Communication of Science and Technology.* London: Routledge.

Burns, T.W., O'Connor, D.J., Stocklmayer, S.M., 2003. "Science Communication: A Contemporary Definition". *Public Understanding of Science* 12 (2), 183–202. https://doi.org/10.1177/09636625030122004

Fahy, D., Nisbet, M. C., 2011. "The Science Journalist Online: Shifting Roles and Emerging Practices". *Journalism* 12 (7), 778–793. https://doi.org/10.1177/1464884911412697

Geipel, A., 2020. *"Don't Act Like a Teacher": How Science YouTubers Become Experts*. Munchen: Munich Center for Technology in Society. https://d-nb.info/1230552812/34

Harney, M., 2003. *Towards a Civic Science – A Mission for the 21st Century*. Dublin: Royal Irish Academy.

Joubert, M., Guenther, L., Metcalfe, J., Riedlinger, M., Chakraborty, A., Gascoigne, T., Schiele, B., Baram-Tsabari, A., Malkov, D., Fattorini, E., Revuelta, G., Barata, G., Riise, J., Schröder, J. T., Horst, M., Kaseje, M., Kirsten, M., Bauer, M. W., Bucchi, M., Flores, N., Wolfson, O., Chen, T., 2023. "'Pandem-icons' — Exploring the Characteristics of Highly Visible Scientists During the Covid-19 Pandemic". *JCOM* 22 (01), A04. https://doi.org/10.22323/2.22010204

Schneider, L., 2017. "Fishy Peer Review at Science by Citizen Scientist Ted Held". www.forbetterscience.com.

Trench, B., 2008. "Towards an Analytical Framework of Science Communication Models". In D. Cheng, M. Claessens, T. Gascoigne, J. Metcalfe, B. Schiele, S. Shi (Eds.), *Communicating Science in Social Contexts – New Models, New Practices*. Dordrecht: Springer, 119–135.

A SHORT LEXICON OF SCIENCE COMMUNICATION

In the academic literature and policy- and practice-oriented discussions of science communication over the past five decades, certain terms have recurred frequently as signposts to paths of analysis and action. Many of these are used in variable forms and meanings and may be fought over on the basis of different usages and interpretations. Understanding the usages of these keywords is crucial to orienting oneself in the increasingly complex field of science communication. We discussed ten of these terms a decade ago (Bucchi and Trench, 2014, 2015). In this Lexicon, we offer an extended list of 15 terms in alphabetical order with short discussions of how they are used and perhaps how best they can be defined. As will be seen, the scope of some terms overlaps, but, even at the risk of some repetition, we have left them stand because the terms are all widely used.

Deficit is a central concept in identifying the intellectual (or ideological) foundations of certain science-in-society ideas and practices and enabling their critique. Two assumptions often underlie this concept: public opinion and political decision-makers are misinformed about science and the issues raised by its development; this misinformation is fuelled by inadequate and sensationalist media coverage of technoscientific topics. This situation is seen as being exacerbated by poor training in basic science and a general lack of interest in scientific research among the institutions and the cultural intelligentsia. From this perception arises the need to propose initiatives covering the gap between experts and the general public, reversing public attitudes towards science and technology or at least attenuating their hostility. Such emphasis on the public's inability to understand the achievements of science has warranted the label of 'deficit model' for this view of the public understanding of science. Critics of the deficit-based approach accept that awareness issues may exist across publics but suggest that this is not the best starting point: focus instead, they say, on what the audiences do know and on their questions. Discussion has continued over many years on what kinds of knowledge about science the public generally lacks and needs to have: knowledge of scientific facts, of scientific theory, of scientific methods or of the organisation and governance of science. A more traditional notion of missing knowledge of facts (see Literacy below) remains widely assumed in contemporary science-in-society practice, notably in contexts where there are perceived problems of pseudoscience and superstition.

Dialogue came to be presented as the acceptable alternative to the deficit model in the late 1990s as science and technology controversies became evident and public demand for involvement in such issues increased. The frequently cited report of the House of Lords Science and Technology Committee in Britain (2000) acknowledged the limits of science communication based on a paternalistic, top-down science-public relationship and detected a 'new mood for dialogue'. In many countries and at the European Commission level, funding schemes and policy documents shifted their keywords from public awareness of science to citizen engagement, from dissemination to dialogue, from science *and* society to science *in* society. The claimed shift from deficit to dialogue remains a powerful narrative in public communication of science, the two approaches being widely seen as distinct and one as inherently superior to the other. Commentaries speak of the 'dialogical turn' as a historical change that has taken effect across Europe and more widely. Dialogue and related approaches are now widely proposed and enacted as alternatives to those defined as deficit-based, particularly in Europe, Australasia and North America. However, it has been argued that some dialogue methods are not genuinely two-way, in that the original sponsors of the communication (generally scientific or policy institutions) stay in control and the citizens taking part have no significant influence on the final outcomes. There is yet another strand to the discussion of communication and cultural practices that draws attention to the possibilities of dialogical events beyond specific political or informational objectives, emphasising instead the process of taking part. Science cafés may be a case in point: the satisfaction for those involved may reside in the exchange itself rather than anything beyond it.

Dissemination has long been the default dominant view of science communication, referring to the spread or diffusion of information and knowledge from science outwards. It is the more generic model of communication that underlies the deficit model (see Deficit above), and equivalent terms are used in several languages in association with popularisation (see Popularisation below). Dissemination is not so tied to science communication as those other terms are; the concept is contained in the word's origins, which refer to the spreading of seed. In the earlier days of science communication, research funders such as the European Commission included dissemination clauses in their contracts: grant recipients were expected to spread the word of their funded work, which typically was done through the construction of a web site, at that time a relative novelty. While the expectations of funders and the aspirations of researchers have developed since then into more interactive formats, the need for dissemination – the spread of relevant information – remains in many circumstances (see Models of communication).

Engagement has become, in many countries, though starting in Britain, a prevalent term for a range of science-in-society practices in policy, education, information or entertainment contexts. Engagement can refer to the actions and attitudes both of knowledge producers and of various sectors of the public. When researchers, for example, go to the streets to talk about their work, this may be called public engagement. The change of vocabulary carries with it, at least implicitly, a shift to an understanding of relations between the partners in the process as more equal and more active. Different levels and modes of engagement are envisaged, for example, by reference to downstream and upstream engagement. The latter has been proposed for priority attention on the basis that early involvement of the public in discussion and eventually negotiation of new developments in science and technology will likely lead to more satisfactory outcomes for all involved, and specifically to knowledge that has earned public trust. The case of genetically modified foods and crops is cited as an example of late, or downstream, public engagement; citizens in many countries across the world were presented with products ready for use and, in many cases, they reacted in a hostile manner. Public engagement activities are nowadays regarded in several countries as a significant dimension of the mandate and responsibility of higher education and research institutions in the context of the so-called 'third mission'.

Expertise is one of the most common forms through which scientific knowledge and actors enter the public domain; scientists take on public roles in validating, interpreting and commenting on developments in science and advising governments and other social institutions on their implications. The pandemic crisis has offered an extraordinary opportunity to observe and study the public role of experts. As producers of knowledge, scientists tend to operate in tightly circumscribed and increasingly specialised spaces. But as experts in public arenas, they are expected to take a broader view and answer media questions or offer policy advice on themes in which they may not be strictly competent. Studies of science in society have often focused on how scientific expertise is expressed and accorded authority in public. Increasingly, expertise of several kinds is involved when complex scientific issues are played out in public arenas. Scientific expertise may be set alongside tacit, informal knowledge that various social groups possess through their experience or culture. Case studies in health and agriculture in the 1980s and 1990s identified as lay expertise the knowledge that patients and farmers brought to a particular issue and that qualified the definition of that topic given by scientific experts. Scientific expertise in contemporary societies is facing challenges arising from greater accessibility of specialist information to non-experts, increasing questioning of the choices and competence of experts and public exposure to controversies among established experts.

Inclusion as a criterion of socially responsible policy and practice is increasingly invoked in discussion of science in society and science communication. Organisations and events are urged to be mindful of whether and how they represent the full spread of society, and specifically minorities – variously defined – that may be marginalised or under-represented. The presence and representation of LGTBQ+, minority ethnic and colonised communities in science and science communication have received particular attention. Inclusion has been the basis of a critique of science communication as reproducing the historical colonialist and hierarchical patterns of science. Social justice and decolonising frames are applied in advocating for a more responsive and inclusive practice of science communication. Coupled with diversity and equality, inclusion is proposed as a prerequisite of practitioner groups and a goal of participation. Inclusion is also applied, though with less emphasis, to consideration of science communication formats. In this respect, a view of science communication as the social conversation around science can be proposed as the most inclusive. In its several usages in relation to practitioners, publics and formats, inclusion is always a continuing work.

Literacy in science or *scientific literacy*, among the general public, was a key concern of science-in-society policy and commentary in the early phases of what became a science communication movement. The measurement of this literacy was the object of the earliest surveys on the public's connection to science. The relevant literacy in question was conceived as knowledge of scientific facts, such as the earth's movement around the sun, and survey respondents were recorded as being able to identify true or false statements on such matters. The concern that motivated such studies was often explicitly stated as one about low levels of literacy; the definition of an acceptable or high level was rarely if ever made explicit. As the limitations of such studies and of their underlying assumptions came to be recognised, a wider discussion grew around the various kinds of literacy in science that might be taken into consideration: knowledge of facts demonstrated by science; knowledge of scientific theories; knowledge of how science works. To these Durant (1994) added: knowledge of how science *really* works. His point was that the picture of how science works presented as adherence to the scientific method was often an idealised picture, very different from the real-life, everyday practice of science. As science-related public surveys have become ever more sophisticated (see Publics below), literacy in the traditional terms outlined here has had greatly reduced emphasis, but it still may be cited in public discourse and in discussions within scientific communities.

Model of communication is one of the key theoretical concepts in science communication. In the early 1990s, sociologists and communication scholars identified problems of theory and conceptualisation in the dominant science practices. They referred, in this context, to the model of communication underlying such practices, meaning the mental construction of relations between the actors in the communication process. They identified the dominant model in terms such as top-down and hierarchical and pointed to the assumption that the target public was defined by a deficit of some kind. Over the past two decades, science communication communities in research and practice have sustained a discussion about the limits of inherited models and the characteristics of models that are more appropriate for the present day. Part of that discussion and research has been explicitly prescriptive and binary: it labels some models of communication as old and discredited and others as new and appropriate. In this context, the shift in preferences from one model to another is represented as evolutionary and irreversible. But another part of that discussion and research is more descriptive and analytical, aimed at understanding better the range of possible models, how different models are applied, how the language used to describe a practice may disguise the model that effectively shapes the practice, how different models can coexist and what governs the choices made.

Participation has come to represent a stronger form of engagement by the public both with scientific ideas and with the governance of science. The term has acquired specific meaning in science in society through association with ideas of participatory democracy and participatory communication. In these contexts, participation implies strongly active citizens who can engage at many levels, including in deliberation on the very topics for negotiation and communication. Thus, participation tends to be used in science in society to refer to an option that goes beyond the deficit-dialogue split and overcomes the need to refer, for example, to 'real dialogue' in order to insist on the authenticity of the process. If deficit and related modes of communication can be considered one-way and dialogue two-way, then participation can be represented as three-way because it implies publics or citizens talking with each other as well as talking back to science and its institutions. Specific versions of public participation in science include 'citizen science' and 'open science'. In the first, citizens may contribute to scientific research as collectors or contributors of data; in the second, researchers make all protocols, data, analyses and publications available online for public scrutiny, allowing the interested public to access science in the making. In some cases, this accessibility paves the way for a substantive contribution, as in the Fold-It project, where the form of certain proteins was identified through open collaboration between experts and non-experts.

Popularisation is the term with the longest tradition among those used to describe a wide range of practices in making scientific content accessible to general, non-expert audiences. The near-equivalent terms in other languages, including vulgarisation (French), divulgazione (Italian) and divulgación (Spanish), also have long and continuing histories and carry similar connotations. In the past century and particularly after World War II, the global and policy landscape redefined popularisation in conceptual and even ideological terms, particularly in the US and Western Europe, with science's social and political role significantly captured by the metaphor of the "goose laying golden eggs", i.e., delivering economic prosperity, social progress and military power if appropriately fed. Popularisation was expected to "sell science" to the broader public to strengthen social support and legitimation. When a new phase of critical reflection on the role of science in development and (more broadly) in society opened, spurred by environmentalist, anti-war and anti-nuclear movements, the concept of popularisation also came under criticism as embodying a paternalistic, diffusionist of science communication. More recent conceptualisations have reappraised the term, considering it suitable to describe communicative interactions, for example, in situations characterised by low public sensitivity or mobilisation, moderate perception of controversy among experts, and great visibility of science actors and institutions involved. In China, popularisation has long been and remains the preferred term to refer to a wide range of science-in-society activities.

Publics has become a common term in discussion and study of science in society, indicating in shorthand that the public is diverse, even fragmented. Because it is not a common, much less everyday, word, publics often has to carry the quote marks around it that draw attention to its deliberate use. Adopting the plural form was an important part of recognising that generalisations about *the* public – specifically in terms of its deficits – are very rarely valid and often seriously misleading. Referring to publics has been associated with the proposal of a contextual model of communication, according to which the communicators inform themselves about and are attentive to, the various understandings, beliefs and attitudes within the public. Beyond the obvious differentiation of publics as young or old, male or female and scientifically educated or not, the plural-publics approach has been supported by the accumulation of data on the widely varying interest, attention and disposition towards scientific matters in the populations of individual countries and, comparatively, across countries and continents. Surveys of public knowledge and public perceptions have become more sophisticated and nuanced, measuring, for example, fine distinctions in levels of trust in scientists and scientific institutions and attitudes towards emerging technologies correlated with educational experiences and world views. A strong focus on publics is almost standard in the training of scientists for public communication; short courses offered to researchers by research councils, universities and professional organisations often start by asking: who are the publics you want to communicate with, and why?

Scientific culture or Culture of Science is used in several variations to refer to the standing of science in the general culture of a country or other cultural context. Two interconnected uses of the term have largely dominated debate in the past few decades; one significantly influenced by the concept of 'two cultures', associated strongly with the chemist–novelist CP Snow (1959), contrasts scientific culture with that of the humanities and the arts and deprecates their separation and the lack of public attention for scientific culture; the second use has been almost interchangeable with public understanding of science in its more traditional and limited meaning. This equates scientific culture with public attention to and interest in scientific topics and levels of scientific literacy and thus, through a deficit perspective, with public acceptance and support of different science and technology developments.

A narrower vision assumes that scientific culture is a distinct and coherent object that can be infused into general culture through appropriate communications. A broader view underscores increasing diversity and fragmentation within science practice; permeability of the boundaries between contemporary science and society; cross-fertilisation between images and narratives in general culture and scientific concepts and ideas; and significant visibility and presence of scientific figures and concepts in the public sphere as well as in contemporary arts. This culture of science in society encompasses not just understanding of specific scientific content but also a social intelligence of science as part of culture. A more technically oriented discussion continues on defining measurements of scientific culture through a combination of traditional indicators in the economy and indicators of science communication activities and of public attitudes to science.

Scientism has been used since the mid-20th century to refer to a dogmatic belief in the superiority of science as a way of knowing the world. Social scientists and philosophers of science identified this critically as a characteristic of scientific communities. It can be seen as an ideology in the sense that it is a conception that has become 'common sense' and widely assumed as valid in those communities. Although it was not generally cited in the original identification of the deficit model of science communication, it appears as a plausible explanation for the emergence and persistence of that approach: scientism has top-down, hierarchical perspectives built-in. It is also an exclusive view of relevant knowledge, downgrading or ignoring the contributions of humanities and social studies. In some earlier uses of the term, scientism referred more neutrally to the scientific way of working, but the distinction between scientific and scientistic – despite the awkwardness of that word – is crucial. In more recent uses of the term, scientism has been adopted in studies of public attitudes to refer to love of or belief in science.

Trust in science has become a central concern of much science communication. Many programmes and policies start from a view of public trust in science as low or too low and set themselves the objective of raising the levels. The prevailing view of low public trust in science among scientists and policymakers has influenced those science communication perspectives: trust in science is presented as a problem that science communication can help fix. In such prescriptive discourses, trust in science is often presented as little different from faith or belief, and mistrust or distrust as 'anti-science'. Much of this echoes wider concerns among national and international leaders about low and declining trust in institutions, in politics, in expertise. But in relation to public trust in science in particular, there is abundant evidence from surveys and other studies that the levels are generally higher than for other professions and institutions, and further, that those levels are stable or increasing. In that context, the continuing insistence on public trust in science as a worsening problem appears as an ideological stance to justify certain kinds of social intervention. Some recent science communication initiatives and studies depart from this kind of prescription, offering nuanced views of types of trust and of the conditions for trustworthiness. This rethinking extends to the recovery of mistrust as a valid view of some science, or of science in some circumstances, equivalent to the scepticism that has been long recognised as an essential part of scientific endeavour.

Visible scientists or public scientists have been present in every generation since modern public figures, but those who did science were not defined as scientists until the 19th century, and up to then the potential public for science was restricted to a shallow layer of the highly educated. With the professionalisation of science, rapid growth in the number of scientists and the development of a mass public, a particular concern grew about the relative invisibility of science to the vast majority of society. The classic American study by Rae Goodell (1977) drew attention to named scientists in psychology, anthropology, molecular biology and other fields who had achieved public visibility as informers and explainers of contemporary science. But it also highlighted institutional constraints, which meant that scientists might as often be punished as rewarded for seeking such visibility. The most successful popularisers exploited the opportunities of the rapidly spreading medium of television to become household names. In astronomy, new technologies and natural history, in particular, photogenic or otherwise charismatic scientists developed highly visible careers as TV presenters. Others became public scientists in myriad ways, as newspaper contributors, TV show panellists, advisory committee or expert group members and as politicians. More recently, the celebrity culture that grew up around entertainment and sport has affected many other sectors, including science. Celebrity scientists' views are sought and broadcast on topics well beyond their areas of recognised expertise, and their private lives become public affairs.

REFERENCES

Bucchi, M., Trench, B., 2014. "Introduction: Science Communication Research: Themes and Challenges". In M. Bucchi, B. Trench (Eds.), *Routledge Handbook of Public Communication of Science and Technology*, 2nd edn. London and New York: Routledge, 1–14.

Bucchi, M., Trench, B., 2015. "Science Communication and Science in Society: A Conceptual Review in Ten Keywords". *Tecnoscienza: Italian Journal of Science & Technology Studies* 7 (2), 151–168.

Durant, J., 1994. "What Is Scientific Literacy?". *European Review* 2 (1), 83–89.

Goodell, R., 1977. *The Visible Scientists*. Boston, Mass: Little Brown.

House of Lords, 2000. *Science and Society: Report of the Selected Committee on Science and Technology*. London: The Stationery Office.

Snow, C. P., 1959. *The Two Cultures*. Cambridge, UK: Cambridge University Press.

RECOMMENDED READING

The following annotated listing of books is offered as a reading list for those undertaking more detailed studies of science communication. The books have been selected on the basis that they present general views of science communication. It is a reflection of the state of the field that the majority of the books are edited collections of contributions, with the variations in depth and breadth that this implies. Only a small number of books present a consistent, worked-out thesis or analysis of the field as a whole.

Bonfadelli, H., Fähnrich, B., Lüthje, C., Milde, J., Rhomberg, M., Schäfer, M., (Eds.), 2017. *Forschungsfeld Wissenschaftskommunikation.* **Wiesbaden: Springer VS.**
(Available in German only) This handbook explores the research field (Forschungsfeld) of science communication through 24 chapters from contributors who come mainly from communication science in the German-speaking world. The collection opens on historical and theoretical issues and moves into sub-fields, including health and risk communication and science journalism, that have developed their own literature and networks.

Broks, P., 2006. *Understanding Popular Science.* **Maidenhead: Open University Press.**
Opening with a historical account of science popularisation from the early 19th century onwards, mainly in a British context, the author develops an analysis of changing science–public relations in the second half of the 20th century, when popular science was redefined.

He advocates an approach based on 'critical understanding of science in public' (CUSP) that emphasises the contexts within which both science and the public exist.

Bucchi, M., 1998. *Science and the Media: Alternative Routes in Scientific Communication.* **London and New York: Routledge.**
Departing from the 'canonical account' of public communication of science, this book examines, through theoretical discussion and case studies, including cold fusion, how and why scientists turn or 'deviate' to the public. It proposes a model of science communication based on concepts drawn from the sociology of science, including paradigm and boundary.

Bucchi, M., Trench, B., (Eds.), 2008, 2014, 2021. *Routledge Handbook of Public Communication of Science and Technology,* **3 editions. London and New York: Routledge.**
This handbook seeks to present the state of the art in science communication studies (as distinct from a handbook as a practical guide). Across its three editions, it has brought together a total of 38 contributors from media and film studies, discourse analysis, sociology of science and of culture, social psychology and other backgrounds, including some of the most influential contributors to the development of the field.

Davies, S. R., Horst, M., 2016. *Science Communication: Culture, Identity and Citizenship.* **New York: Palgrave Macmillan.**
The authors define science communication as an ecosystem and apply tools from cultural studies to describe that system. They examine the roles of various agencies, notably those of 'academic capitalism' and public relations. They expand the range of formats and contexts to be considered in analysing and practising science communication, including, for example, images and emotions, but emphasising throughout the dimensions of democracy and citizenship.

Dierkes, M., von Grote, C, (Eds.), 2000. *Between Understanding and Trust: the Public, Science and Technology.* **London: Routledge.**
These essays scrutinise two key ideas in public and policy discourse around science, understanding and trust, including the editors' provocative question, Why should the public 'understand' science? US and British policy contexts are examined, and survey data from the European Union, the United States, Japan and Canada used in a cross-country comparison of scientific literacy and attitudes toward science and technology.

Erickson, M., 2005, 2015. *Science, Culture and Society: Understanding Science in the 21ˢᵗ Century*, 2 **editions. Cambridge, UK: Polity.**
Through selected cases and issues, the author aims to show how science is made, or 'constructed', in society and culture, involving many parts of society and not just scientists. The examples and illustrations come from a wide range of scientific and cultural activities, and the theoretical discussion covers key ideas in philosophy, history and sociology of science. It provides context for science communication but also addresses directly popular science and science fiction.

Fage-Butler, A., Ledderer, L., Hvidtfelt Nielsen, K., (Eds.), 2025. *Science Communication and Trust.* **Singapore: Palgrave Macmillan.**
A wide-ranging examination of relations between science communication and trust through 24 chapters from perspectives in science studies, communication studies, philosophy, rhetoric and social psychology, among other disciplines. Often within the context of the COVID-19 pandemic, the widely assumed possibilities and purposes of science communication in increasing public trust in science are analysed critically from many perspectives, as are the many conceptualisations of trust and the relations of trust to knowledge, beliefs and emotions.

Fahy, D., 2015. *The New Celebrity Scientists: Out of the Lab and into the Limelight.* **Lanham, USA, and London: Rowman and Littlefield.**
Eight case studies serve to show how publicly visible scientists become celebrities and, with that, their personal opinions and even private lives become public matters. The careers of people such as Richard Dawkins and Stephen Hawking are laid out to demonstrate how such scientists contributed to their moulding by the media into celebrities, alongside those drawn from other spheres, and thus acquired exceptional powers to circulate scientific ideas in popular culture.

Felt, U., Davies, S. R., (Eds.), 2020. *Exploring Science Communication: a Science and Technology Studies Approach.* **London: Sage.**
The editors open and close the collection with discussion of conceptual issues in applying STS to science communication, stating their aim to encourage a conversation between the two fields. The STS approach presented here is close to cultural studies, considering science communication to be not representational but generative of

new meanings. Eight case studies cover visual climate communication, public space, public health and emotion, among other topics.

Fox Keller, E., 2003. *Making Sense of Life: Explaining Biological Development with Models, Metaphors, and Machines.* **Cambridge, USA: Harvard University Press.**
A theoretical physicist turned science historian and philosopher asks what counts as an explanation of biological development and identifies the use of physical and mathematical models and of mechanical and computing-derived metaphors in molecular biology as ways of understanding life. The book focuses on representations, including images, in scientific communication but raises important issues also for public communication.

Frankel, F., 2018. *Picturing Science and Engineering.* **Cambridge, USA: MIT Press.**
Research scientist in chemical engineering and science photographer Felice Frankel offers a richly illustrated guide for creating science images that are both accurate and visually stunning, starting with advice to researchers to step back from their own notions of visualisation to try to imagine how others outside their field might perceive it and to consider the appropriate metaphor or analogy to inform the image.

Friedman, S. M., Dunwoody, S., Rogers, C. L., (Eds.), 1999. *Communicating Uncertainty: Media Coverage of New and Controversial Science.* **London and Mahwah, New Jersey: Lawrence Erlbaum Associates.**
This collection of essays focuses on uncertainty, controversy and complexity in science and in its media representation. The editors and contributors include leading figures in the practice and study in the US of media science. Each of the three sections closes with an account of a round table discussion, reflecting and extending the diversity of views around the topics.

Gregory, J., Miller, S., 1998. *Science in Public: Communication, Culture and Credibility.* **London and New York: Plenum Press.**
One of the earliest attempts to bring together knowledge and ideas around science communication for a general readership, it remains a reference text over a quarter-century later. Written by two British authors in the context of the public understanding of the science movement that emerged in Britain in the 1990s, it raises critical questions about that approach.

Holliman, R., Whitelegg, E., Scanlon, S., Thomas, J., (Eds.), 2009. *Investigating Science Communication in the Information Age: Implications for Public Engagement and Popular Media.* **Oxford: Oxford University Press.**
One of two collections of essays arising from the Open University's education programme in science communication – the other being practice-focused – contains chapters by mostly British-based scholars in science studies or media studies that explore issues in dialogue, journalism, popular science and audiences.

Irwin, A., Wynne, B., (Eds.), 1996. *Misunderstanding Science? The Public Reconstruction of Science and Technology.* **Cambridge, UK: Cambridge University Press.**
British social scientists examine in these essays the relations between experts and publics around particular science-related topics and arenas in the context of the then-current 'public understanding of science' approach. The editors' Introduction and Conclusions provide still-relevant context and critique as they seek to shift focus from the public misunderstanding of science to scientific institutions' misunderstanding of the contexts in which they work.

Leßmöllmann, A., Dascal, M., Gloning, T., (Eds.), 2020. *Science Communication.* **Berlin: De Gruyter.**
This collection of essays includes arguments for applying semiotics, rhetorics, speech communication, visual communication, discourse and terminological analysis and other humanities approaches to science communication. Historical or "evolutionary" perspectives are strongly represented here as a means of situating the topic in context. The concept of medialisation is analysed critically, and another chapter tracks how increasingly complex models of science communication have developed.

Lévy-Leblond, J-M., 1996. *La Pierre de Touche: la science à l'épreuve.* **Paris: Folio Essais, Gallimard.**
(Available in French only) This is one of several collections of essays by a theoretical physicist who has been a prolific writer and editor on science in society and culture. In short pieces originally published in magazines and journals, the author explores issues of language, politics and history in science and argues for science's *mise-en-culture* (placing into culture), including through science criticism.

Nelkin, D., 1987, 1995. *Selling Science: How the Press Covers Science and Technology*, 2 editions. **New York: Freeman and Company.**

Sociologist Nelkin provided a foundation for studies of media science with this critical account of US newspapers' and news magazines' coverage of selected issues. It draws attention to common characteristics and structural constraints of science coverage and to the risks of hype, but also adversarial relations between scientists and journalists.

Priest, S., Goodwin, J., Dahlstrom, M. F., (Eds.), 2018. *Ethics and Practice in Science Communication.* **Chicago: University of Chicago Press.**
This collection explores an aspect of science communication that has only in recent years received concentrated attention. The book does not propose a code of ethics but considers the ideas and principles that would inform such a code. The case studies refer to public issues such as climate change, agricultural biotechnology and industry environmental campaigns, raising questions of responsibilities and duties and the challenges associated with communicating controversies and risks.

Rödder, S., Franzen, M. Weingart, P., (Eds.), 2012. *The Sciences' Media Connection: Public Communication and Its Repercussions.* **Dordrecht: Springer.**
Seven of the nineteen chapters focus on the concept of medialisation include theoretical reflections by co-editor Weingart, one of the first to adopt the concept in relation to science communication. Other topics include news values, framing and story types in media coverage of science. Three chapters offer a 'practitioner's perspective'.

Schiele, B., Claessens, M., Shi, S., (Eds.), 2012. *Science Communication in the World.* **Dordrecht: Springer.**
Contributors come from a wide geographical spread, including from Latin America, Asia (India, Korea and China), Australia, Canada and several European countries, as well as the cross-country perspective of the European Commission. Their topics include models, historical development, recent trends and prospects of science communication in these various contexts. 'Horizontal' issues include science communication education, science culture and visible scientists.

Wilkinson, C., Weitkamp, E., 2016. *Creative Research Communication: Theory and Practice.* **Manchester: Manchester University Press.**
Two science communication educators review the possibilities for innovation and invention in communicating about research across various contexts such as digital formats, art and politics. The book

adopts the term research communication as wider than science, and it combines discussion, examples and practical guidance.

Bucchi, M., Trench, B., (Eds.), 2016. *Public Communication of Science.* **London: Routledge.**

This anthology in four thematic volumes is intended to provide an overview of the research effort around this topic and to be a library purchase and reference book. The studies and essays were selected as representing the most significant contributions to science communication studies, including historical pieces that set the foundations for this field of work. Many of the readings here have been referenced in earlier chapters of this book. These materials are also available in their original settings, as indicated in the following tables of contents.

Volume I: Theories and Models

1. Ludwik Fleck, *Genesis and Development of a Scientific Fact.* (Chicago University Press, 1970), pp. 98–125.
2. C. P. Snow, 'The Rede Lecture 1959: Part 1, The Two Cultures', *The Two Cultures* (Cambridge University Press, 1959), pp. 1–21.
3. Robert K. Merton, 'The Matthew Effect in Science', *Science*, 1968, 159, 3810, 56–63.
4. Leon Trachtman, 'The Public Understanding of Science Effort: A Critique', *Science Technology and Human Values*, 1981, 6, 36, 10–15.
5. Thomas F. Gieryn, 'Boundary-Work and the Demarcation of Science from Non-Science: Strains and Interests in Professional Ideologies of Scientists', *American Sociological Review*, 1983, 48, 781–95.
6. Michel Cloitre and Terry Shinn, 'Expository Practice: Social, Cognitive and Epistemological Linkages', in T. Shinn and R. Whitley (Eds.), *Expository Science: Forms and Functions of Popularisation. Sociology of the Sciences*, IX (D. Reidel Publishing Company, 1985), pp. 31–60.
7. Bruno Latour, 'Literature', *Science in Action: Following Scientists and Engineers Through Society* (Harvard University Press, 1987), pp. 21–44.
8. Christopher Dornan, 'Some Problems in Conceptualising the Issue of "Science and the Media"', *Critical Studies in Media Communication*, 1990, 7, 1, 48–71.
9. Leah A. Lievrouw, 'Communication and the Social Representation of Scientific Knowledge', *Critical Studies in Mass Communication*, 1990, 7, 1, 1–10.

10. Stephen Hilgartner, 'The Dominant View of Popularization: Conceptual Problems, Political Uses', *Social Studies of Science*, 1990, 20, 3, 519–39.

11. Brian Wynne, 'Knowledges in Context', *Science, Technology and Human Values*, 1991, 16, 1, 111–21.

12. Mike Michael, 'Lay Discourses of Science: Science-in-General, Science-in-Particular, and Self', *Science, Technology and Human Values*, 1992, 17, 3, 313–33.

13. Jean-Marc Lévy-Leblond, 'About Misunderstandings about Misunderstandings', *Public Understanding of Science*, 1992, 1, 1, 17–21.

14. Baudouin Jurdant, 'Popularization of Science as the Autobiography of Science', *Public Understanding of Science*, 1993, 2, 4, 365–73.

15. Massimiano Bucchi, 'When Scientists Turn to the Public', *Public Understanding of Science*, 1996, 5, 4, 375–94.

16. Peter Weingart, 'Science and the Media', *Research Policy*, 1998, 27, 8, 869–79.

17. Steven Miller, 'Public Understanding of Science at the Crossroads', *Public Understanding of Science*, 2001, 10, 1, 115–20.

18. Massimiano Bucchi, 'Can Genetics Help Us Rethink Communication? Public Communication of Science as a "Double Helix"', *New Genetics and Society*, 2004, 23, 3 269–83.

19. Maja Horst, 'In Search of Dialogue: Staging Science Communication in Consensus Conferences', in D. Cheng, M. Claessens, T. Gascoigne, J. Metcalfe, B. Schiele, and S. Shi (Eds.), *Communicating Science in Social Contexts: New Models, New Practices* (Springer, 2008), pp. 259–74.

20. Brian Trench, 'Towards an Analytical Framework of Science Communication Models', in D. Cheng, M. Claessens, T. Gascoigne, J. Metcalfe, B. Schiele, and S. Shi (Eds.), *Communicating Science in Social Contexts: New Models, New Practices* (Springer, 2008), pp. 119–38.

21. Peter Weingart, 'The Lure of the Mass Media and Its Repercussions on Science', in P. Weingart, M. Franzen, and S. Rödder (Eds.), *The Sciences' Media Connection – Public Communication and Its Repercussions: Sociology of the Sciences Yearbook*, 28 (Springer, 2012), pp. 17–32.

22. Alan Irwin, 'Risk, Science and Public Communication: Third-Order Thinking about Scientific Culture', in M. Bucchi and B. Trench (Eds.), *Handbook of Public Communication of Science and Technology*, revised edn. (Routledge, 2014), pp. 160–72.

Volume II: Processes and Strategies

23. J. B. S. Haldane, 'How to Write a Popular Scientific Article', in K. Dronamraju (Ed.), *What I Require from Life: Writings on Science and Life from J. B. S. Haldane* [1941] (Oxford University Press, 2009), pp. 154–60.

24. Peter Medawar, 'Is the Scientific Paper a Fraud?', in P. Medawar, *The Threat and the Glory: Reflections on Science and Scientists* (Oxford University Press, 1963), pp. 228–33.

25. Rae Goodell, 'What's a Nice Scientist Doing in a Place Like the Press?', *The Visible Scientists* (Little, Brown, 1977), pp. 120–41.

26. Royal Society, *The Public Understanding of Science, Sections 2, 3, 5, 6, and 7* (The Royal Society, 1985), pp. 9–16, 21–8.

27. Rae Goodell, 'How to Kill a Controversy: The Case of Recombinant DNA', in S. Friedman, S. Dunwoody, and C. L. Rogers (Eds.), *Scientists and Journalists: Reporting Science as News* (The Free Press, 1986), pp. 170–81.

28. Susan L. Star and James R. Griesemer, 'Institutional Ecology, "Translations" and Boundary Objects: Amateurs and Professionals in Berkeley's Museum of Vertebrate Zoology 1907–39', *Social Studies of Science*, 1989, 19, 387–419.

29. David M. Phillips, E. J. Kanter, B. Bednarczyk, and P. L. Tastad, 'Importance of the Lay Press in the Transmission of Medical Knowledge to the Scientific Community', *New England Journal of Medicine*, 1991, 325, 16, 1180–3.

30. Anders Hansen, 'Journalistic Practices and Science Reporting in the British Press', *Public Understanding of Science*, 1992, 3, 2, 111–34.

31. Bruce V. Lewenstein, 'From Fax to Facts: Communication in the Cold Fusion Saga', *Social Studies of Science*, 1995, 25, 3, 403–36.

32. Jean-Marc Lévy-Leblond, 'The Case for Science Criticism', *La Pierre de Touche: La Science a l'épreuve* (Editions Gallimard, 1996), pp. 149–64 (a new translation by David Denby).

33. Hans Peter Peters, 'The Interaction of Journalists and Scientific Experts: Co-operation and Conflict between Two Professional Cultures', *Media Culture and Society*, 1995, 17, 1, 31–48.

34. Carl Sagan, 'No Such Thing as a Dumb Question', *The Demon-Haunted World: Science as a Candle in the Dark* (Headline, 1996), pp. 300–17.

35. Tom Wilkie, 'Sources in Science: Who Can We Trust?', *Lancet*, 1996, 347, 1308–11.

36. Tim Radford, 'Science for People Who Don't Want to Know About Science', *Accountability in Research*, 1997, 5, 39–43.

37. Jane Gregory and Steven Miller, 'ABC of Risk: Apples, Beef, and Comets', *Science in Public: Communication, Culture and Credibility* (Plenum Press, 1998), pp. 166–95.

38. Sharon Dunwoody, 'Scientists, Journalists and the Meaning of Uncertainty', in S. M. Friedman, S. Dunwoody, and C. L. Rogers (Eds.), *Communicating Uncertainty: Media Coverage of New and Controversial Science* (Lawrence Erlbaum Associates, 1999), pp. 59–79.

39. Stephen Jay Gould, 'Preface', *The Lying Stones of Marrakech: Penultimate Reflections in Natural History* (Jonathan Cape, 2000), pp. 1–3.

40. Alan Irwin, 'Constructing the Scientific Citizen: Science and Democracy in the Biosciences', *Public Understanding of Science*, 2001, 10, 1, 1–18.

41. Susanna Hornig Priest, 'Re-Inventing Milk', *A Grain of Truth: The Media, the Public, and Biotechnology* (Rowman and Littlefield, 2001), pp. 17–34.

42. Vincent Kiernan, 'Diffusion of News about Research', *Science Communication*, 2003, 25, 1

43. Hans Peter Peters, Dominique Brossard, Suzanne de Cheveigné, Sharon Dunwoody, Monika Kallfass, Steve Miller, and Shoji Tsuchida. 'Science-Media Interface: It's Time to Reconsider', *Science Communication*, 2008, 30, 2, 266–276.

44. Stuart Allan, 'Making Science Newsworthy: Exploring the Conventions of Science Journalism', in R. Holliman, E. Whitelegg, E. Scanlon, S. Smidt, and J. Thomas (Eds.), *Investigating Science Communication in the Information Age: Implications for Public Engagement and Popular Media* (Oxford University Press, 2008), pp. 149–65.

45. Brian Trench, 'Science Reporting in the Electronic Embrace of the Internet', in R. Holliman, E. Whitelegg, E. Scanlon, S. Smidt, and J. Thomas (Eds.), *Investigating Science Communication in the Information Age: Implications for Public Engagement and Popular Media* (Oxford University Press, 2008), pp. 166–79.

46. Declan Fahy and Matthew Nisbet, 'The Science Journalist Online: Shifting Roles and Emerging Practices', *Journalism*, 2011, 12, 7, 778–93.

47. Bernard Schiele, 'Science Museums and Centres: Evolution and Contemporary Trends', in M. Bucchi and B. Trench (Eds.), *Handbook of Public Communication of Science and Technology*, revised edition (Routledge, 2014), pp. 40–57.

Volume III: Publics for Science

48. Helga Nowotny, 'Experts and Their Expertise: On the Changing Relationship between Experts and Their Public', *Bulletin of Science, Technology and Society*, 1981, 1, 2, 235–241.

49. Jon D. Miller, 'Scientific Literacy: A Conceptual and Empirical Review', *Daedalus*, 1983, 112, 2, 29–48.

50. Maurice Goldsmith, 'The Proper Public for Science', *The Science Critic: A Critical Analysis of the Popular Presentation of Science* (Routledge and Kegan Paul, 1986), pp. 1–16.

51. Steven Shapin, 'Science and the Public', in R. C. Olby et al. (Eds.), *Companion to the History of Modern Science* (Routledge, 1990), pp. 990–1007.

52. Brian Wynne, 'Misunderstood Misunderstanding: Social Identities and Public Uptake of Science', *Public Understanding of Science*, 1992, 1, 281–304.

53. John Durant, 'What Is Scientific Literacy?', *European Review*, 1994, 2, 1, 83–9.

54. Geoffrey A. Evans and John R. Durant, 'The Relationship Between Knowledge and Attitudes in the Public Understanding of Science in Britain', *Public Understanding of Science*, 1995, 4, 1, 57–74.

55. Steven Epstein, 'The Construction of Lay Expertise: AIDS, Activism and the Forging of Credibility in the Reform of Clinical Trials', *Science, Technology and Human Values*, 1995, 20, 4, 408–37.

56. Michel Callon, 'The Role of Lay People in the Production and Dissemination of Scientific Knowledge', *Science, Technology and Society*, 1999, 4, 1, 81–94.

57. Benoit Godin and Yves Gingras, 'What Is Scientific and Technological Culture and How Is It Measured? A Multidimensional Model', *Public Understanding of Science*, 2000, 9, 1, 43.

58. Bernadette Bensaudé-Vincent, 'A Genealogy of the Increasing Gap Between Science and the Public', *Public Understanding of Science*, 2001, 10, 99–113.

59. Sheila Jasanoff, 'Technologies of Humility: Citizen Participation in Governing Science', *Minerva*, 2003, 41, 223–44.

60. Dietram A. Scheufele and Bruce V. Lewenstein, 'The Public and Nanotechnology: How Citizens Make Sense of Emerging Technologies', *Journal of Nanoparticle Research*, 2005, 7, 659–67.

61. Martin Bauer, Nick Allum and Steven Miller, 'What Can We Learn from 25 Years of PUS Research? Liberating and Expanding the Agenda', *Public Understanding of Science*, 2007, 16, 79–95.

62. Ulrike Felt and Maximilian Fochler, 'The Bottom-up Meanings of the Concept of Public Participation in Science and Technology', *Science and Public Policy*, 2008, 35, 7, 489–99.

63. Susanna Hornig Priest, 'Reinterpreting the Audiences of Media Messages about Science', in R. Holliman, E. Whitelegg, E. Scanlon, S. Smidt, and J. Thomas (Eds.), *Investigating Science Communication in the Information Age: Implications for Public Engagement and Popular Media* (Oxford University Press, 2008), pp. 223–36.

64. Edna F. Einsiedel, 'Publics and Their Participation in Science and Technology: Changing Roles, Blurring Boundaries', in M. Bucchi and B. Trench (Eds.), *Routledge Handbook of Public Communication of Science and Technology* (Routledge, 2014), pp. 125–39.

Volume IV: Media Representations of Science

65. Allan Mazur, 'Media Coverage and Public Opinion on Scientific Controversies', *Journal of Communication*, 1981, 31, 2, 106–15.

66. Marcel C. LaFollette, 'Science on Television: Influences and Strategies', *Daedalus*, 1982, 111, 4, 183–97.

67. Roger Silverstone, 'Narrative Strategies in Television Science: A Case Study', *Media, Culture & Society*, 1984, 6, 4, 377–410.

68. Harry Collins, 'Certainty and the Public Understanding of Science: Science on Television', *Social Studies of Science*, 1987, 17, 689–713.

69. Daniel Jacobi and Bernard Schiele, 'Scientific Imagery and Popularized Imagery', *Social Studies of Sciences*, 1989, 19, 731–53.

70. Ulrike Felt, 'Fabricating Scientific Success Stories', *Public Understanding of Science*, 1993, 2, 375–90.

71. Dorothy Nelkin, 'Promotional Metaphors and Their Popular Appeal', *Public Understanding of Science*, 1994, 3, 25–31.

72. Dorothy Nelkin, 'The Scientific Mystique', *Selling Science: How the Press Covers Science and Technology*, 2nd edn. (W. H. Freeman, 1995), pp. 14–30.

73. Suzanne De Cheveigné and Eliseo Veron, 'Science on TV: Forms and Receptions of Science Programmes on French Television', *Public Understanding of Science*, 1996, 5, 231–53.

74. Matthew C. Nisbet and Bruce V. Lewenstein, 'Biotechnology and the American Media: The Policy Process and the Elite Press, 1970 to 1999', *Science Communication*, 2002, 23, 4, 359–91.

75. Brigitte Nerlich, Robert Dingwall, and David D. Clarke, 'The Book of Life: How the Completion of the Human Genome Project Was Revealed to the Public', *Health: An Interdisciplinary*

Journal for the Social Study of Health, Illness and Medicine, 2002, 6, 4, 445–69.

76. Esa Väliverronen, 'Expert, Healer, Reassurer, Hero and Prophet: Framing Genetics and Medical Scientists in Television News', *New Genetics and Society*, 2006, 25, 3, 233–48.

77. Martin W. Bauer, Kristina Petkova, Pepka Boyadjieva, and Galin Gornev, 'Long-Term Trends in the Public Representation of Science across the "Iron Curtain": 1946–1995', *Social Studies of Science*, 2006, 36, 1, 99–131.

78. Anabela Carvalho, 'Ideological Cultures and Media Discourses on Scientific Knowledge: Re-reading News on Climate Change', *Public Understanding of Science*, 2007, 16, 223–43.

79. M. Schäfer, 'From Public Understanding to Public Engagement: An Empirical Assessment of Changes in Science Coverage', *Science Communication*, 2009, 30, 4, 475–505.

INDEX

For Product Safety Concerns and Information please contact our EU
representative GPSR@taylorandfrancis.com
Taylor & Francis Verlag GmbH, Kaufingerstraße 24, 80331 München, Germany